THE CHESS MYSTERIES OF
SHERLOCK HOLMES

50
Tantalizing Problems
of **Chess Detection**

Raymond M. Smullyan

DOVER PUBLICATIONS, INC.
MINEOLA, NEW YORK

"Which Color?," the problem on p. 21,
was first published in slightly different form
in *Scientific American,* April 1974.

Bibliographical Note

This Dover edition, first published in 2012, is an unabridged
republication of the work originally published by Alfred A. Knopf,
Inc., New York, in 1979.

International Standard Book Number

ISBN-13: 978-0-486-48201-9
ISBN-10: 0-486-48201-4

Manufactured in the United States by Courier Corporation
48201403
www.doverpublications.com

To My Wife BLANCHE

and to the Memory

of My Brother EMILE

and of My Dear Friend

THEODORE SHEDLOVSKY

CONTENTS

CONTENTS

ACKNOWLEDGMENTS

First, I wish to thank a graduate student of my Princeton days, who went through earlier versions of several of these puzzles, and provided a host of helpful suggestions. For years I have tried to recall his name but have unfortunately failed. I hope he will see this and get in touch with me, so I can thank him by name in my next book of chess puzzles.

R.M.S.

A NOTE FOR
THE CHESS DETECTIVE

Suppose I told you that in the following position no pawn has ever reached the eighth square. Would you believe me?

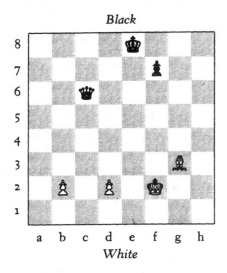

Black

White

If you did, you shouldn't have, because what I have told you is logically impossible! Here is the reason why.

To begin with, in all the problems in this book, squares will be designated by letter and number. For example, in the above position, the White king is on f2, the Black king is on e8, a White bishop is on g3, the Black queen is on c6, and White pawns are on b2 and d2.

Now, how did the White bishop ever get to g3 from its home square of c1, since the pawns on b2 and d2 have never moved to let it out? The only possibility is that the bishop originally on c1 was captured on its home square without

having moved, and that the bishop on g3 is really a promoted bishop. (After all, a pawn doesn't have to promote to
a queen; it can also promote to a rook, a bishop, or a
knight). Therefore, the statement that no pawn has ever
reached the eighth square simply doesn't hold water!

The above problem, like the one on the jacket cover, is a
very simple example of the type considered by Sherlock
Holmes in this remarkable manuscript. Such problems belong to the field known as retrograde analysis. Unlike the
more conventional type of chess problem (which is concerned with the number of moves in which White can win),
these problems are concerned only with the *past* history of a
game. The variety of questions that these puzzles can pose is
quite fascinating. For example, you might exhibit a position
in which one of the pieces is dropped (or represented by a
coin lying on the square), and the problem is to figure out
what the piece is. Then again, positions are given from
which you can deduce that one of the pieces on the board is
a promoted piece, but it is impossible to tell which piece it
is. (Indeed, a position is given in which you cannot even determine whether the promoted piece is White or Black!)

It is even possible, as we shall see, to prove that White
has a mate in two moves from a certain position, while at
the same time it is *impossible* to show the mate! Unbeliev-
able as this may sound, it is true.

These problems are intriguing studies in pure deductive
reasoning. They might be said to lie on the borderline between logic and chess (in fact, they have sometimes been referred to as problems in *chess-logic*). They very much have
the psychological flavor of detective stories, and naturally
had an enormous appeal for Holmes—indeed, this is the
only type of chess problem in which he took any interest.
We are most fortunate in that Holmes's brilliant exposition
of this whole subject in Part I is so lucid that any reader who
merely knows how the pieces move will easily be able to follow his explanations step by step. He will have become
pretty much of an expert in this type of reasoning by the

time he reaches Part II, and will be adequately prepared to help Holmes locate Captain Marston's buried treasure by means of retrograde analysis and, at the same time, solve a curious double-murder mystery.

It is our great good luck that Holmes was so adept at this type of chess problem. If he hadn't been able to solve one of them in particular (you will find out which one), this manuscript would never have taken shape, for he would have fallen prey to a diabolical scheme of Moriarty's and lost his life before he ever even met Dr. Watson.

RAYMOND M. SMULLYAN

Elka Park, New York
February 1979

THE CHESS MYSTERIES OF
SHERLOCK
HOLMES

50
Tantalizing Problems
of **Chess Detection**

PART I

SHERLOCK HOLMES
at the
CHESSBOARD

A MATTER
OF DIRECTION

Whathat about a stroll to the chess club?" Holmes remarked one early afternoon.

"Why, Holmes!" I cried in amazement. "I did not know you were a chess enthusiast!"

"Not of the conventional sort," laughed Holmes. "I do not have too much interest in chess as a *game*—indeed, I do not have much inclination for games in general."

"But what is chess, if *not* a game?" I asked in astonishment.

Holmes's face grew serious. "There are occasional chess situations, Watson, which challenge the analytic mind as fully as any which arise in real life. Moreover, I have found them as valuable as any exercises I know in developing those powers of pure deduction so essential to dealing with real-life situations."

"Tell me more," I replied with interest.

"What I have in mind, Watson, is this: In an actual game, both players have their eyes fixed entirely on the future. Each player tries to control the future in a way favorable to his own position. Also, in most chess *problems* of the usual sort—White to play and mate in so many moves—the entire emphasis is on doing something to control the future. Now, although I have the deepest respect for the better problems of this sort—many of them are really ingenious works of art!—the type of strategies involved, clever as they are, is hardly of any use to me in my own work."

"I am afraid I am still in the dark," I responded.

"There are certain chessboard situations," explained

Holmes, "which are of no interest to the *player* of chess as a game—of no interest with regard to future outcomes—but are of vital interest in providing clues as to what must have happened in the *past*."

"Can you give me an example, Holmes?" I asked with ever-growing curiosity.

"Another time," said Holmes, rising. "Right now I really do feel like taking a jaunt to the chess club. Why don't you come with me, Watson? Who knows—we might encounter an actual situation to illustrate my point."

I thought this a good idea and got my hat, and together we sauntered over to the club. It was empty except for two occupants: Colonel Marston, whom we knew fairly well, and a distinguished, intelligent-looking gentleman with a very pleasant and humorous manner.

"Why, Holmes," said Marston, rising from his place at the chessboard, "let me introduce you and Dr. Watson to a very dear friend of mine, Sir Reginald Owen. We have just finished a most delightfully bizarre and eccentric game. The playing was utterly wild on both sides, though perfectly legal, of course."

"So I see," remarked Holmes, looking at the board. The position was this:

North

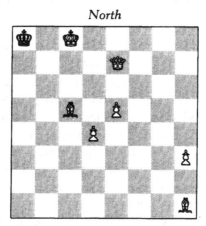

South

"Why is it, Marston," said Holmes, "that every time I see you at chess, you always seem to be playing White?"

Marston laughed at this, but suddenly his face dropped. "Why, Holmes," he said, "how on earth did you know I was playing White? I'm sure the last move was made some time *before* you or Watson entered the room. So how could you possibly know?"

"Perhaps," said Sir Reginald, with a slightly mischievous smile, "your friend knew for the simple reason that you have a White piece clenched in your hand."

"I don't see that as evidence," I protested. "Why could it not be a piece captured from his opponent?"

"As a matter of fact," laughed Holmes, "Sir Reginald happens to have an excellent point. I have observed that when a chess player is idly fondling a piece during the course of a game, nine times out of ten it is one of his own men."

"Why is that?" I asked.

"That's difficult to say, Watson; I think it is a matter of unconscious courtesy. One has a certain natural psychological reluctance to handle one's opponent's pieces. At any rate, the fact is that players almost always handle their own pieces. However, that was not my clue, since Colonel Marston opened his hand only *after* Sir Reginald's remark, hence I did not see what color piece Marston was holding. But even if I had, my inference would have been only a probability, and I would not have told Marston as confidently as I did that he was playing White. In this case, my inference was far more definite; I knew for *certain* that Marston was playing White."

"But how?" pleaded Colonel Marston.

Holmes replied, with a merry twinkle in his eye, "I really think, sir, you will have more fun trying to solve this little mystery yourselves. I'm sure to bump into you again one of these days, Marston, and if you have not figured it out by then, I shall be delighted to tell you. But I'm afraid you will find it disappointingly elementary!"

Both Marston and Sir Reginald became intensely absorbed in the problem and started discussing with each other all sorts of wild possibilities. After a short while we took our leave, after shaking hands with our new acquaintance, Sir Reginald.

"Holmes," I said when we were outside, "I am bursting with curiosity. How *did* you know?"

"We can discuss that quietly this evening over a chessboard. Meanwhile, what say you to a visit to the Museum, and then a good dinner at Agosino's?"

♟

Several hours later, we were comfortably back at Baker Street, Holmes lounging in his favorite dressing gown and smoking his well-loved pipe. "Is it not a delightful coincidence," he chuckled, "that so soon after our rather abstract conversation this afternoon, we encountered a situation amenable to retrograde analysis?"

"Retrograde analysis?" I asked. "What is that?"

"Why, the very thing we were discussing! Have you yet figured out how I knew Marston was playing White?"

"Why, no," I replied. "I used all the methods you have taught me, I examined the whole room thoroughly, but could not discover a single clue!"

At this, Holmes burst into a roar of laughter. "The whole room, Watson, the whole *room!* Did you examine the rest of the building as well?"

"I never thought of that," I admitted meekly.

"My dear Watson," Holmes said, laughing harder than ever, "I was only jesting, you know. It was hardly necessary to examine the whole building, nor even the whole room, nor the table, nor the players, but merely the chessboard."

"The chessboard? What was peculiar about the chessboard?"

"Why, the position, Watson, the *position*. Don't you recall anything peculiar about the position?"

"Yes, I do recall that at the time I regarded the position as highly unusual, but I cannot see how one can deduce that Marston played White from it!"

At this Sherlock rose. "Let us set up the position again. . . . Now there," said Holmes after reconstructing the setup of the afternoon, "can't you deduce which side is White and which side is Black?"

I looked long and carefully, but could discover no clue at all. "Is this an example of what you call 'retrograde analysis'?" I asked.

"A perfect example," replied Holmes, "albeit an extremely elementary one. But come now, you see no clue whatsoever?"

"None at all," I said sadly. "Superficially it would appear that White is on the South side. But this is really quite superficial! The game is clearly in the end-game stage, where it is not too uncommon for one of the kings to be driven to the opposite end of the board. So it seems that White could really be on either side."

"There is *nothing* in the situation which arrests your attention, Watson?" asked Holmes despairingly.

I looked again at the board. "Well, Holmes, I suppose there is one feature which would probably arrest anybody's attention, namely that the Black king is now in check from the White bishop. But I can't see that this has any bearing on which side is White."

Holmes smiled triumphantly. "All the bearing in the world, Watson. And here is where retrograde analysis comes in! In retro-analysis—which is short for 'retrograde analysis'—one must delve into the past. Yes, the *past*, Watson! Since Black is now in check, what could have been White's last move?"

I looked again at the board and replied, "Why, it could easily have been the White pawn on e5 just having moved from e4 and discovering check from the bishop. This, of course, assumes that White is South. But on the other hand, it could also be that White is North, in which case his last

move was the pawn on d4 from d5. I see no basis for deciding between these two possibilities."

"Very good, Watson, but if it is really true, as you have
said, that White's last move was with one of the pawns on
e5 or d4, then what could have been Black's move immediately before that?"

I looked again and replied, "Obviously by the Black king,
since it is the only Black piece on the board. He couldn't
have moved from b8 or b7, hence he must have moved out
of check from a7."

"Impossible!" cried Holmes. "Had he been on a7, he
would have been simultaneously in check from the White
queen and the other White bishop on c5. If the queen had
moved last to administer check, Black would have already
been in check from the bishop. Had the bishop moved last,
Black would have already been in check from the queen.
Such an impossible check is technically known in retrograde
analysis as an 'imaginary check.' "

I thought for a moment, and realized that Holmes
was right. "Then," I exclaimed, "the position is simply
impossible!"

"Not at all," laughed Holmes. "You have simply not considered all the possibilities."

"Now look, Holmes, you yourself have just proved that
Black had no possible last move!"

"I proved nothing of the sort, Watson!"

At this point I was getting a bit impatient. "Oh, come
now, Holmes, you just proved to my entire satisfaction that
the Black king had no possible last move."

"True enough, Watson, I proved that the Black *king* had
no possible last move, but this hardly proves that *Black* had
no possible last move."

"But," I cried, "the king is the *only* Black piece on the
board!"

"The only Black piece on the board *now*," corrected
Holmes, "but that does not mean that the king was the only

Black piece on the board immediately prior to White's last move!"

"Of course," I replied, "how stupid of me! White on his last move could have *captured* a Black piece. But," I exclaimed, more puzzled than ever, "whichever of the pawns on e5 or d4 moved last made no capture!"

"Which only proves," laughed Holmes, "that your original conjecture that White's last move was with one of those two pawns is simply incorrect."

"Incorrect!" I cried, bewildered. "How can that be?" Then it dawned on me! "Of course!" I exclaimed triumphantly. "I see the whole thing now! How silly I did not see it all along! White's last move was with the pawn from g2 capturing a Black piece on h3. This capture simultaneously checked the Black king and captured a Black piece, and it was this piece—whatever it was—which made the preceding Black move!"

"Nice try, Watson, but I'm afraid it won't do! If a White pawn had just been on g2, then how on earth could the bishop on h1 ever have gotten to that square?"

Here was a new puzzler! At this point I said, "Really, Holmes, I am now thoroughly convinced that the position is simply impossible!"

"Really now? Well, well! This only affords another example of what I have often remarked: Conviction, no matter how firm, is not always a guarantee of truth."

"But we have exhausted *every* possibility!" I exclaimed.

"All but one, Watson—the *correct* one, as it happens."

"It seems to me that we have really covered every possibility. I am certain that we have *proved* this position to be impossible!"

Holmes's expression grew grave. "Logic," he replied, "is a most delicate—a most fragile—thing. Powerful as it is when used correctly, the least deviation from strict reasoning can produce the most disastrous consequences. You say you can 'prove' the position impossible. I should like you to try to give me a *completely rigorous* proof of this fact. I think that in so doing, you may yourself discover your own fallacy."

"Very well, then," I agreed, "let us review the possibilities one by one. We—or rather you—have certainly proved that neither pawn on d4 or e5 could have moved last. Correct?"

"Absolutely," said Holmes.

"Likewise the pawn on h3?"

"Right," said Holmes.

"Surely the bishop on h1 didn't move last!"

"Right again," said Holmes.

"And certainly the other bishop on c5, and the White queen, could not have moved last. And *surely* the White king didn't move last!"

"I'm completely with you so far," remarked Holmes.

"Well then," I said, "the proof is complete! *No* White piece could have moved last!"

"Wrong!" exclaimed Holmes triumphantly. "That is a complete non sequitur!"

"Just a minute now," I cried, a bit beside myself. "I have accounted for *every* White piece on the board!"

"Yes," said Holmes, highly amused at my consternation, "but not for pieces *off* the board."

At this point, I was beginning to doubt my sanity. "Really now, Holmes," I cried in utter desperation, "since White moved last, the piece he just moved must be on the board, since Black has not yet moved to capture it. Pieces don't just move off the board by themselves, you know!"

"Wrong," said Holmes, "and there lies your whole fallacy!"

At this point, I blinked my eyes and shook myself to convince myself that I was really awake. With the utmost control, I calmly and slowly said, "You honestly mean to tell me, Holmes, that in chess a piece may leave the board without being captured?"

"Yes," replied Holmes. "There is one and only one type of piece which can do that."

"A pawn!" I said, with a profound sigh of relief. "Of course, a pawn on reaching the eighth square promotes. But," I continued, "I do not see how that can help us in the present situation, since the White queen is not now on the eighth square—regardless of which direction White is going."

Holmes replied: "Is there any rule in chess which demands that a pawn, when it promotes, must promote to a queen?"

"No," I replied. "It can promote to a queen, rook, bishop, or knight. But how does that help us here? . . . Hallo," I said, "of course! It may have promoted to the bishop on h1—which of course means that White is North. But how does that leave a last move for Black? Ah, I've got it! The promoting White pawn was on g2 and *captured* a Black piece on h1; this Black piece made the move right before that! So indeed, White must be North!"

"Very good, Watson," said Holmes with a calm smile.

"One thing, though, that bothers me, Holmes: Why on earth should White have promoted to a bishop when he could have had another queen?"

"Watson," Holmes very carefully replied, "that question belongs to psychology and probability, and certainly not to retrograde analysis, which deals not with probabilities, but only with absolute certainties. We never assume that a player has played *well*, but only that he has played *legally*. So however improbable it is that a given move was made, if no other move was possible, then that must have been the move which in fact *was* made. As I have told you many times, when one has eliminated the impossible, then whatever remains, however improbable, must be the truth."

A DELIGHTFUL
VARIATION

If I have gone to excessive lengths in spelling out every last detail of the preceding adventure, I have done so for the sake of the reader who is an absolute beginner in the art of chess detection. This was the stage I was in at the time. As I grew more and more adept at this type of reasoning, Holmes could afford to give me shorter and shorter explanations of the solutions to these intriguing problems.

Two evenings later, Holmes and I had a pleasant surprise: an unexpected visit from Colonel Marston and our new acquaintance, Sir Reginald. We were always delighted to see Colonel Marston, who is a keen-witted and most amusing chap. And Sir Reginald we liked more and more as we got to know him better.

After a cozy round of warm brandy, Marston said, "It did not take us very long, Holmes, to figure out your little mystification. Of course, since *we* were the players, this is not very much to our credit. After all, Holmes, it was *I* who made that highly unorthodox move of promoting to a bishop. But it took us a bit of thought to realize that from the standpoint of that position, no other last move could have been possible."

Holmes, delighted, arose and said, "I knew you would get it, Marston, I knew you would! And now, gentlemen, I should like to tell you of a curious coincidence. The position which I found when you were playing Sir Reginald bears a striking similarity to the following position I saw two years ago in a club I was visiting in Calcutta." And Holmes set up the chessboard as follows:

North

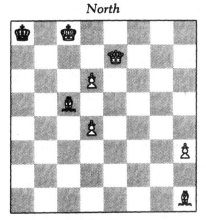

South

"As you see, gentlemen, the only difference is that a pawn is on d6 instead of e5. Yet this little difference is enough to make it totally impossible to tell now which side is White."

We all looked at the board with great curiosity. Marston spoke first. "I really don't see how this change is relevant, Holmes. White's only possible last move is a pawn from g2 capturing a Black piece on h1 and converting to a bishop."

"Not so, Marston; this is *a* possible last move, but not the only one. When I first saw this position, I had the same immediate thought, and was just about to surprise the players by telling them which side was White when I suddenly realized that there was another equally real possibility. Fortunately, just as one of the players was about to clear the board, I asked him, 'Sir, has any promotion taken place in this game?' His hands stopped as he turned to me and said, 'No, there hasn't, but why do you ask?' 'Because,' I replied triumphantly, 'this means that you must have been playing White!' Both gentlemen looked at me rather amazed, though their amazement turned to amusement when I told them how I knew."

We all looked at the position with greater interest than ever. Marston again spoke. "I can't see it, Holmes; I really can't! Since White's last move was not a promotion, it must have been with one of the pawns on d4 or d6. But in either case, Black could not have had any possible last move!"

"Not so, Marston, not so," replied Holmes. "White's last move was indeed with one of these pawns. I'll even give you a hint and tell you that it was with the pawn on d6."

"But," said Marston, "the pawn must have moved from d5 to give Black check from the bishop on h1. So how could the Black king have moved before that? The only square it could have come from is a7, but this would involve an imaginary check!"

At this point Sir Reginald spoke up. "Brilliant, Mr. Holmes, really brilliant! I see it now! White moved the pawn not from d5, but from e5, capturing a Black pawn on d5 *en passant!* This Black pawn had, of course, just moved from d7. White's move before that was with the pawn then on e5 from e4, putting Black in check. And Black's move before *that* was with the king from a7, moving out of check from the bishop on c5, but *not* being in check from the queen, since the Black pawn would have been on e7."

"I'm afraid that was a bit hard for me to follow," I remarked. "Would you mind going over it again?"

"Why certainly," replied Sir Reginald. "The position a couple of moves ago was something like this." He rearranged the pieces:

Then he made the following sequence of moves to bring the game to its present position:

1	B – c5 check	K – a8
2	P – e5 check	P – d4
3	P x P en passant check	

And that's how Holmes knew which side was which!

A LITTLE
EXERCISE

Several days later, I had a memorable evening alone with Holmes, during which I learned more about retro-analysis than perhaps on any other occasion. I was becoming vastly intrigued with this subject, and I began by asking, "Holmes, do all retrograde problems involve a final checkmate position?"

"Oh, not at all," he replied. "Most of them, as it happens, do not." I then asked, "And is the question always which side is which?"

"Most definitely not," replied Holmes. "This question happens to be exceedingly rare! Here, let me set up a little exercise to illustrate the more normal type of situation:

Black-1

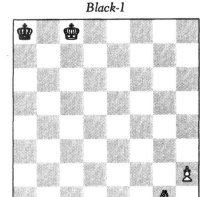

White-3

"I call this an 'exercise,' Watson, since it is really too simple to dignify by the word 'problem.'

"As you see, Watson, neither side is mated—nor even in check. And we are given that your side is White. The question now is this: Given that Black moved last, what was his last move, and White's last move?"

I thought for a while, then said, "Holmes, I'm sorry to be such a slow pupil, but the situation again seems impossible! Obviously Black just moved out of check from a7, but I don't see how White could possibly have moved his bishop to administer the check!"

"Not bad, Watson; not bad at all! I see you are beginning to think. But why do you have this persistent habit of forgetting that a move may involve a capture?"

Then, of course, I saw it. "Right, Holmes, right. Black's last move was with the king from a7 *capturing* a White piece on a8. This piece must have moved before that out of the diagonal from g1 to a7 to uncover check from the bishop. What piece could that be? Why obviously a knight, which had moved from b6 to a8. Thus Black's last move was from a7 to a8, capturing a White knight."

"Correct," said Holmes.

A new thought suddenly occurred to me. "Holmes," I said, "was it really necessary in this problem to be given which side was White?"

"Of course," replied Holmes. "If we hadn't been given that information, then a second solution would have been possible: A White pawn could have just promoted to bishop."

A few minutes later I asked, "Holmes, are all retrograde problems solved by first considering what was the last move?"

"Oh no," replied Holmes. "In many of these problems there is no way at all of ascertaining the last move. Nevertheless one can determine a move—or sequence of moves—

which have occurred *sometime* earlier in the game, though precisely when they occurred is neither determinable nor relevant to the solution of the problem."

"Can you give me an example?" I asked. Holmes thought for a moment and then set up the following position:

Black-14

White-12

"The problem is: On what square was the White queen captured?"

I looked at the position, and reasoned thus: "Well, Holmes, I see that White is missing his queen, both bishops, and one knight. Now, two captures can be accounted for by the Black pawns on e6 and h6; the first came from d7 and made a capture on e6, and the second came from g7 and made a capture on h6. Now, neither White bishop ever got out onto the board to be captured by these pawns, since the one from c1 was hemmed in by the pawns on b2 and d2 which have not yet moved, and the one from f1 was hemmed in by the pawns on e2 and g2."

"Good, Watson," interrupted Holmes. "I'm glad you were able to see that yourself."

"Elementary, my dear Holmes," I could not help but jest. "However, I'm afraid my reason cannot carry me much further. All right, we now see that the White pieces captured on e6 and h6 are the queen and a knight. So the White queen was captured either on e6 or h6, but I cannot see why the queen could not have been captured on either one of them and the knight on the other."

"Well," said Holmes, "then I'll have to give you some hints in the form of questions. One of the main things in solving these problems is to think of the right questions to ask oneself. Now, what was captured by the White pawn on b3?"

"Obviously a Black bishop," I replied.

"What is the home square of that bishop?"

"Clearly c8, as the bishop from f8 travels only on black squares."

"Right. Now comes the crucial question: Which was captured first, the Black bishop or the White queen?"

"I see no way to tell," I replied.

"Well then, put it this way: Did the White queen get captured before or after the capture on b3 by the White pawn?"

I looked at the position again, and began to see the point. "The White queen," I said, "got out onto the board via the square a2, hence the pawn on b3 made its capture *first* to let the queen out. And since the pawn captured a bishop, then the bishop was captured before the queen."

"Exactly," said Holmes. "And now, does this not solve the problem?"

"I do not see how."

"Well then, I guess the next question to ask yourself is this: Did the bishop from c8 get captured on b3 *before* or *after* the capture on e6?"

"The pawn on e6 made its capture *first*," I replied, "to let the bishop out."

"Correct," said Holmes. "Now you have all the pieces of the puzzle, Watson; you have merely to put them together."

"Ah," I said, "now I see it. The capture on e6 was made before the Black bishop got out to be captured on b3, which in turn happened *before* the White queen got out to be captured. Therefore, the White queen was *not* the piece captured on e6. In other words, the sequence was this: First the knight was captured on e6, then the Black bishop got out and was captured on b3, then the White queen got out, and must have been the piece captured on h6. So the queen was captured on h6."

"Very good, Watson," said Holmes encouragingly. "I think with a bit more experience, you will be able to do retrograde analysis."

WHICH COLOR?

In retrograde analysis," I asked Holmes one day, "is the problem always to ascertain either the direction, the last move, or the square on which a given piece was captured?"

"Oh, by no means," replied Holmes. "The questions that arise are sometimes of a far more intriguing and bizarre nature."

"Can you give me an example?" I inquired.

Holmes thought for a moment. "Not a sufficiently elementary one for you to comprehend at your present stage. . . . But wait! I have just thought of an ideal problem for you right now!"

With a curious look of amusement, Holmes set some pieces on the board. "Wait, Watson, the position is not complete!" said Holmes as he went over to his shelf and brought back a tiny box which had often caught my eye. "I've many times had the feeling, Watson, that you were on the verge of asking me what is in this box, but politeness forbade you. Am I right?"

"Why yes, Holmes," I confessed, "I've many times wondered what is in that box."

"Well now, Watson, let us see!"

With the air of a mischievous child, or of a stage magician, he slowly and rather melodramatically opened the lid and removed, of all things, a pawn. The pawn was identical in shape to the pawns of Holmes's chess set, only this pawn was painted half black and half white.

"Where on earth did you get that?" I laughed.

"Oh, Watson, I keep this as a memento of the little problem I am about to show you." So saying, Holmes placed the mysterious pawn on g3, and the position was this:

Black-1 or 2

White-3 or 4

"In this game," said Holmes, "no piece or pawn has ever moved from a white square to a black square, nor from a black square to a white square. The question is: What is the color of the pawn on g3?"

"That's easy," I jested. "The pawn on g3 is Black and White!"

"No, seriously," laughed Holmes, "what color pawn— Black or White—should be put on g3 to make the position compatible with the given conditions?"

I looked, but could see no clue at all. A thought struck me: "You say that no piece has ever moved from a square of one color to a square of another. Under these circumstances, how could the knights have ever moved?"

"Good point, Watson, good point! Obviously, they never did!"

"Then whatever happened to them?"

"Clearly they were captured on their own squares."

This struck me as a trifle odd, to say the least, but as certainly within the realm of possibility. "Is this a hint to the solution?" I asked.

"No, Watson, not to this problem—though it does figure in some other problems of this type."

"Then I am completely in the dark; I see no clue whatsoever."

"The clue," Holmes remarked, "is in the position of the White king."

"The White king!" I exclaimed. "The White king could have come from any of the four squares a3, a5, c5, or c3."

"Yes, yes," said Holmes, "but how did he ever escape from his home square, e1?"

I looked at the position, perplexed. The king could not have escaped via the black squares d2 or f2, since the pawns there had not yet moved. And, of course, he could not have escaped via the squares d1, e2, or f1, since they are white. So how did he get out? Then I saw it: "White castled," I said triumphantly.

"Excellent, Watson! Did he castle on the king's side or the queen's side?"

I thought for a moment and said, "On the king's side."

"Why?" asked Holmes.

"Because had he castled on the queen's side, the queen's rook would have had to move from a1—a black square—to d1—a white square."

"Better and better," cried Holmes. "Now, does that not solve the mystery of our little pawn?"

"I don't see how, Holmes," I replied, but then I suddenly saw it: "Of course! The only way the king could get from g1 to his present position is via the squares h2 and g3. If the pawn on g3 is White, then it must have come from h2, and the king could never have got out! So the mystery pawn must be Black."

"Now, Watson," he said, "you are really beginning to get the hang of retrograde analysis!"

ANOTHER
MONOCHROMATIC

These monochromatic problems, Watson—by which I mean problems of the type we just considered, in which it is given that no piece has moved from a white square to a black square, or vice versa—these monochromatics can sometimes lead to the most bizarre situations imaginable. Here is one, Watson, in which the very question is a bit of a shock!"

Black-1

White-3

"In this game," Holmes went on, "no piece has moved from a white square to a black one, and vice versa. Furthermore, the White king has made under fourteen moves. Prove that a promotion has taken place."

I gasped with amazement. "Really, Holmes, really now, this is too much!"

Holmes was delighted with my consternation. "No, Watson, this problem really has a solution."

"Give me *some* clue," I begged.

"I have already given you one," remarked Holmes.

"When?" I asked, quite puzzled.

"In the *last* problem," he remarked. "Recall our conversation about what happened to the knights."

I shall not give you a detailed account of the dialogue that followed, but simply tell you the solution that Holmes extracted from me with his series of Socratic questions.

The four missing knights were captured on their home squares. Then what pieces captured them? There is no problem at all accounting for the White knights nor the Black knight from g8. The only problem is the Black knight from b8. What White piece could have captured it? It could not be the queen from d1, as she travels only on white squares. It could not be the bishop from c1, since the pawns on b2 and d2 have not moved to let it out. It certainly could not be the bishop from f1, because that bishop travels on white squares. Therefore, if the knight on b8 was captured by a queen or bishop, it must have been captured by a *promoted* queen or bishop. If the knight was captured by a *pawn*, then the pawn must have promoted! Now, the knight was not captured by the White king, since it would take the king at least fourteen moves to get there and back. And, of course, it was not captured by a knight, as knights can't move. The last possibility to consider is a rook. Why couldn't the Black knight have been captured by the rook from a1? This is the loveliest part of the problem. The reason is that the rooks—since they are confined each to one color—can go only an *even* number of squares forward, backward, or sideways. In particular, the rooks from a1 and h1 can never possibly be on the second, fourth, sixth, or eighth rows! Therefore, if the Black knight was captured by a rook, then it was captured by a *promoted* rook. This exhausts all possibilities, and we see that each of them involves the promotion of a White pawn. Q.E.D.

A QUESTION
OF SURVIVAL

The last problem that Holmes showed me that day was another monochromatic. "Before we leave these monochromatics, Watson," he said, "there is one more I wish you to consider. It is unique in many respects. The solution is not at all 'complex'—indeed, there is only one simple idea behind it. But this idea cannot be arrived at by any complicated process of ratiocination; it can only be grasped by a single act of intuition."

Full of curiosity, I watched as Holmes laid out the following position:

Black-1

White-4

A White bishop was placed equally between the squares e3 and e4. Thinking this was an oversight, I was about to move it, when Holmes stopped me: "No, no, Watson!

That's precisely the problem! On which square, e3 or e4, stands the bishop—given that this game was monochromatic?"

I looked down helplessly, and Holmes continued speaking, evidently enjoying his own words. "The beautiful thing about this problem," he said rhapsodically, "is that it is so delightfully abstract! Indeed, if it will help you any, I could just as well have had a Black bishop there instead of a White one. Or I could have used a pawn, a rook, or a queen instead of a bishop; it would have made no difference! And," he went on, waxing more and more enthusiastic, "instead of using those two particular squares, I could have used any two adjacent ones. In fact, they needn't even be adjacent— they need only be of different colors! In fact," he added triumphantly, "I could even take the bishop off the board and restate the problem thus: In this monochromatic game, there is another piece somewhere on the board. Does it stand on a white square or a black square?"

Far from being "helped" by this information, I felt more benighted than ever! "Tell you what, Watson," Holmes continued a bit mischievously, "I'll give you a hint! When I was a boy, someone once told me the story of the lion and the bear. A lion and a bear were fighting furiously and devouring each other. Finally each had eaten the other up, and there was nothing left of either!"

"That's a *hint*?" I asked in astonishment.

Holmes, who perhaps had not heard me, continued, "Although I was quite young at the time, I had a sufficient commonsensical grasp of the law of conservation of matter to realize the absurdity of the situation!"

"And that's a *hint*?" I repeated.

"Why yes, Watson, if you look at it right! Don't you see," Holmes continued, with unusual intensity, "that if the bishop were on the wrong color, we would have that exact same impossibility? Look, suppose the bishop were on a white square. Then what piece on the black squares could have captured the last piece which fell on a black square?

Certainly not the White king or the pawns, since they have never moved! Here, put the matter this way: Think of an army of Black pieces on black squares and an army of White pieces on black squares 'devouring' each other, so to speak. There must be at least *one* survivor! What could this survivor be? Only the bishop! Hence it is on a black square."

I was lost in admiration. "Really, Holmes, this is the most remarkable problem you have shown me yet! Who invented this masterpiece?"

"Moriarty," was the thunderously unexpected reply.

"Good God, no!" I gasped.

"Oh yes, Watson! And it is hardly surprising, you know. This problem has the sort of diabolical simplicity which was so much a part of Moriarty's nature."

We sat several minutes in silence. I was recalling those last dramatic days culminating in Holmes's death struggle with Moriarty over the chasm. Holmes, evidently reading my thoughts, said, "Yes indeed, Watson, it seemed as if the lion and the bear were about to devour each other, and that nothing would remain. . . . Yet," he added with an unusual smile, "the lion appears to have survived."

MYSTERY OF
THE MISSING PIECE

Holmes," I said one day, "do you remember that when you first introduced me to retrograde analysis, you said that these problems vary enormously, ranging from the very elementary to some of great complexity?"

"Yes, I remember well," replied Holmes.

"Well then, how about trying me on a difficult one?"

Holmes looked at me with an expression of pleasant surprise. "Watson, your enthusiasm is admirable—most admirable. But really, my boy, you must learn to walk before you can run, you know!"

As fate would have it, I had not long to wait before I saw an astounding demonstration of Holmes's ability to analyze a vastly complicated situation. It happened completely unexpectedly and quite spontaneously, as follows.

Two days later, I arose rather late and found Holmes at breakfast with a letter in hand. "Good morning, Watson," he said as he tossed it to me. "Would you care to go?"

I perused the letter. It was an invitation from Sir Reginald to attend what he called an "informal gathering" to be held at his estate in Surrey a couple of evenings hence. He apologized for the lateness of the invitation, but stressed the "informality" of the occasion and concluded by saying: "So if you and Dr. Watson would care to drop in casually, we would be delighted to have you."

We went. The gathering, when we arrived, proved to be far larger than anticipated, and we saw little of Sir Reginald that evening. But we had the following delightful adventure, in which Holmes gave that astonishing demonstration which I shall never forget.

I was particularly anxious to see Sir Reginald's library, as I had lately learned that our host had an unusual and renowned rare-book collection, and I am somewhat of a bibliophile. When Holmes and I entered this room, there were only two other occupants—two very distinguished-looking gentlemen at the far corner of the library, deeply engrossed in a game of chess.

"May we intrude?" asked Holmes as we walked towards the game.

"Why, certainly," replied one of the gentlemen pleasantly as he lifted his head. When we got to the board—or rather, I should say, chess table—the position was this:

Black-10 or 11

White-10 or 11

On h4 there rested a shilling instead of a chess piece. "May I inquire," I said, "why you are using a coin instead of a chess piece?"

"Why, Sir Reginald's children were playing with the set early this afternoon," replied the same gentleman, "and the youngest absentmindedly ran off with one of the pieces and misplaced it somewhere in this enormous, rambling house.

I understand the servants have been looking for it for hours, but it still has not turned up."

"Doubtless it will," said Holmes. "And what piece is it?"

To my surprise, at this point the other gentleman—clearly the senior of the two—arose and said, "Are you not Mr. Sherlock Holmes?"

"Why, yes," replied my friend.

"And this, I presume, is your friend Dr. Watson?"

"Yes," I acknowledged.

"How delightful to meet you," he said with genuine enthusiasm. "Permit me to introduce myself—Arthur Palmerston, and my brother, Robert. We are both great admirers of yours, you know!"

"How did you know I was Holmes?" inquired my friend.

"No mystery; no mystery at all," laughed Mr. Palmerston. "Under the circumstances one need not be a Sherlock Holmes to recognize Sherlock Holmes."

"Under the *circumstances?*" inquired Holmes.

"Mr. Holmes," he continued, "do you recall the artist Joseph Adler?"

Holmes knit his brows thoughtfully for a moment. "Yes," he recollected. "I met him once in a pub about six months ago. Mr. Adler is a caricaturist, I believe?"

"Oh yes, indeed!" replied Palmerston. "And a most excellent one. Yesterday at lunch, the conversation turned in your direction. He took an old piece of brown paper and on the spot drew a caricature of you the likes of which I have never seen! I hope I will have the pleasure one day of showing it to you, Mr. Holmes. Despite an exaggeration of the features, the *inner* likeness is really fantastic! I think I would have recognized you anywhere in the world!"

"Ah!" said Holmes. "And now, Mr. Palmerston, you have still not told me which of the chess pieces is missing."

Upon which Robert Palmerston jestingly replied, "If you are truly Sherlock Holmes, *you* should be able to tell *us* which is the missing piece!"

Holmes, quite unoffended, good-naturedly replied, "That is hardly a fair test, sir. If this position is such that it is logically deducible what piece lies on that square, then I am sure I would be able to deduce it. But, you know, the odds are about a hundred to one against finding a position arrived at in actual play which is amenable to retrograde analysis."

"In that case," replied the elder Palmerston, "I'll be delighted to wager you a hundred shillings to one that you can't tell us what the missing piece is."

Holmes thought for a moment. "Actually, gentlemen, that bet is quite unfair—unfair to *you*, that is. I could simply *guess* what the piece is, and since I would guess aright with a probability of one out of ten, and you are offering me odds of a hundred to one, then of course I would take the bet, except that it would not be fair of me to do so."

Palmerston corrected himself and said, "What I should have said, Mr. Holmes, is that I will wager you a hundred to one that you cannot *know* what this piece is."

"And how can you know if I *know?*" inquired Holmes.

"I mean to say," continued Palmerston, "that I will wager you a hundred to one that you cannot tell us correctly what the piece is, and *prove* that it must be this piece."

Holmes thought for a moment, then replied, "I am not usually a betting man, Mr. Palmerston, but I must say that this challenge intrigues me. Very well, I accept! Only how much time will you permit me?"

"How much time would you like?" inquired Palmerston.

"Would half an hour be reasonable?" asked Holmes.

"Perfectly," Palmerston replied. "As a matter of fact, my brother and I were both getting a bit stiff sitting so long at the game, and I think a half-hour jaunt on the grounds outside would do us both a bit of good. When we return, Mr. Holmes, I assure you I will be delighted if you win the wager!" So saying, he left the library, followed by his brother.

When both Palmerstons had gone, Holmes at once began studying the position with the eye of an eager hawk. The problem seemed too difficult for me, so I withdrew and started browsing among the books. And what a library it was! The glowing reports I had received were no exaggeration. For example, there was a first edition of the *Characteristics* by the Right Honourable Anthony Cooper, Earl of Shaftesbury, completed in 1710. This was a book long out of print, and exceedingly rare in *any* edition. I settled myself comfortably for a good half-hour of reading. Just fifteen minutes later, however, Holmes's voice rang out triumphantly: "I've got it, Watson, I've got it! I can't wait for the gentlemen to return!"

As it happened, they returned not fifteen, but only five minutes afterwards. "Don't be alarmed, Holmes," said the senior Palmerston, "you still have ten minutes. Or would our presence disturb you?"

"Not at all, gentlemen," Holmes replied, "though as a matter of fact, I have already solved your little problem."

"Really!" exclaimed Palmerston in genuinely happy astonishment. "That is amazing! Robert and I were discussing the situation outside. As you must now realize, Holmes, we both played wildly and eccentrically—though, of course, perfectly legally. We both came to the conclusion, though, that we had pretty well covered up our strange tracks, and that you could never find them!"

"Only the minor ones," replied Holmes. "The major tracks are still quite visible. And they are all that are needed to solve this mystery."

With eager anticipation, the three of us sat down at the chess table; Holmes stood lecturing us, much in the manner of a professor addressing three university students.

"The first clue," began Holmes, "is obviously that Black is in check from the rook on d7. How did White administer this check? This stumped me for a minute, until I realized that White's last move must have been a pawn from c7 capturing a Black piece on d8 and promoting to a White rook."

"Yes," grinned Robert Palmerston, "that was one of my eccentric moves."

Holmes continued, "Well then, the next question I considered was what Black piece on d8 had the White pawn captured? It couldn't be a rook, since that rook would be checking the White king, and would have had no square to come from to give this check, nor could any White piece—even if the unknown were White—have moved from any of the squares e8, f8, or g8, to uncover this check. Therefore, the Black piece just captured on d8 was not a rook. Similarly it could not be a queen."

"Just a moment," I interrupted thoughtlessly. "Why couldn't it be a Black queen having come in diagonally—say from b6?"

"Because," said Holmes, "before the capture, a White pawn was on c7." I felt a bit foolish, and resolved not to interrupt again. "Therefore," continued Holmes, "the piece must have been a knight or bishop."

"Just a moment," I said, quite forgetting my resolution. "How could it be a Black knight when there are already two Black knights on the board?"

"Really now, Watson," said Holmes with a trace of irritation, "why do you always forget the possibility of an underpromotion?" Feeling more foolish than ever, I resolved again to just listen.

"Of course it could have been a knight," said Holmes. "Only if it was, then either it or one of the other two Black knights must have been promoted." For some unaccountable reason, Arthur Palmerston blanched a little at this. "On the other hand," continued Holmes, "if it was a bishop, then it must have been a promoted bishop."

At this point I again could not contain myself. "Why?" I asked.

"Because," Holmes replied, this time more gently, "the original bishop from f8 has fallen on its own square, since neither of the pawns on e7 and g7 has yet moved to let it out."

"Oh, of course," I replied.

"Well then," continued Holmes, "I now realized that some time earlier in the game, a Black pawn had also underpromoted."

"This is amazing," said Arthur Palmerston. "I did indeed underpromote—and this was one of *my* eccentric moves. Robert and I predicted that you would realize his underpromotion, but I am amazed that you found clues leading to mine!"

"They are elementary," said Holmes, "really elementary. The interesting part of the analysis is yet to come. Now then, at this point I was able to deduce that the missing piece must be White."

"How?" inquired Robert Palmerston.

"Well, it couldn't be a Black rook or queen, as then both kings would now be simultaneously in check."

"I see."

"It couldn't be a Black pawn, since only one Black pawn is missing, and we know that it promoted to a bishop or knight. Finally, it couldn't be another Black knight, nor the bishop from f8, nor, of course, another promoted bishop, since this would involve an extra promotion, and only one Black pawn is missing. Therefore the unknown piece must be White."

"Brilliant," said Robert Palmerston.

"Now comes the hard part," said Holmes. "What White piece is it? It obviously can't be a pawn, since the only missing White pawn has promoted to a rook. Therefore the unknown is a queen, rook, bishop, or knight, but which? I first tried eliminating possibilities one by one, but was unsuccessful. At this point, I began sadly to suspect that the problem might be unsolvable. But then suddenly I had an idea! If it worked—which it did—I would be able to eliminate three of the four possibilities in one fell swoop!

"I first asked myself which of the Black pawns did the promotion. That was no problem; the pawn on a6 came from b7, the pawns on c5 and d6 must have collectively come from c7 and d7, hence the pawn on c4 must have come from f7. Therefore the missing pawn comes from h7. Next, I asked, on which square did the missing pawn promote? This was also quite easy; it could not have marched straight down the h-file, since the White pawn on h2 has not yet moved. Hence the Black pawn made at least one capture. It could not have made more than one capture, since there are eleven White pieces on the board, including the unknown on h4, which we now know to be White. Thus five White pieces are missing. One of them was captured by the pawn on a6; three more of them were captured by the pawn on c4 coming from f7. This accounts for four of the missing five, hence the pawn from h7 could not have captured more than one. Thus, this pawn captured *exactly* one piece, and it promoted on g1."

"You know," interrupted Arthur Palmerston, "it really gives one the most uncanny feeling to have one's past moves traced so accurately by such inflexible logic!"

"Well," laughed Holmes, "the end is almost in sight, and now comes the most delicate part of the entire analysis! For some reason I suddenly became intrigued by the following question: We know that the Black pawn from h7 captured a White piece *somewhere* on the g-file, but on what square? Superficially it would appear to be on g2 *behind* the White pawn on g3. But is this necessarily true? Could not the pawn on g3 really have come from f2, leaving the g-file wide open for the advancing Black pawn? Well now, suppose the pawn on g3 really came from f2, then it made one capture. That means the pawn on d8—now in the form of a promoted rook—must have come from g2 having made five captures; four to get from g2 to c6, and one more getting from c7 to d8. So, this possibility would involve a total of six captures of Black pieces. Now, there are exactly six Black pieces missing, so at first glance this would seem possible!"

"Why shouldn't it be possible?" I asked.

"Ah, Watson," replied Holmes almost fiendishly, "we

must recall that the Black bishop from f8 never got out onto the board, but was captured on its own square!"

"Of course," I replied.

"So, gentlemen, this possibility would involve just *one* capture too many! Therefore the pawn on g3 did *not* come from g2. So the Black promoting pawn really *did* capture on g2."

"Why do you make so much of this detail?" I asked. "Is it really relevant?"

"Relevant!" said Holmes, almost in a shout. "Relevant! Why, it solves the entire problem," he exclaimed triumphantly.

"How?" I asked.

"Because g2 is a *white* square. Thus the promoting Black pawn captured on a white square. And the pawn on a6 made *its* capture on a white square. And the pawn on c4 made all its three captures on white squares. Thus all five missing White pieces were captured on white squares. Now the one White piece which cannot be captured on a white square is the bishop from c1. If the unknown on h4 is not this bishop, then this bishop would have to be one of the five missing pieces to be captured on a white square. This is impossible. Hence, gentlemen, your mystery piece is a White bishop."

We all three sat in dumfounded silence. This was one of the most amazing pieces of logical deduction I had ever seen. Surprising as all this was to me, I can well imagine the eerie effect it must have had on the players who all this time *knew* it was a bishop. And by a delightful little flourish of fate, at that very moment, the butler solemnly entered the room holding the White bishop in his hand, which he placed on the table, announcing: "The missing piece has been found, gentlemen."

YOU REALLY CAN'T, YOU KNOW!

A month later we received another invitation from Sir Reginald. We accepted with pleasure. The gathering was far smaller than on the preceding occasion.

Sir Reginald led us into the library. To our delight, there were the Palmerston brothers, again at a game of chess. Just as we entered, I saw Arthur Palmerston completing a move with a White knight. Then they both saw us and jumped up to greet us. After the usual formalities, we all sat down for a delightful evening.

"I'm sorry I missed your fantastic demonstration last month," said Sir Reginald. "But, of course, Arthur and Robert have explained it to me down to the last detail. Really quite brilliant, Holmes!"

"As a matter of fact," said Holmes with an extremely grave expression, "I was really silly—most silly about the entire thing! When I got home, I suddenly realized that I could have solved the whole mystery far more efficiently than I did! Yes indeed, had I used empirical rather than purely logical methods, I could have solved the problem in a fraction of the time!"

"How?" asked both brothers almost simultaneously.

"Why," said Holmes, with his old merry twinkle, "by opening the box of chess pieces which you gentlemen so trustingly left behind when you went out onto the terrace."

This got a good general laugh. The conversation then turned to the question of whether Holmes, had he used such an unfair method, would have technically won the bet. Sir Reginald maintained that he would not, arguing that the terms of the bet specified that Holmes not only *name* the piece correctly, but *prove* it was the right one. Arthur Palmerston maintained that Holmes would have won.

"It all depends on how you define the word 'proof,'" he said. "After all, proofs are of two kinds—deductive and inductive. The word 'deductive' did not enter into the terms of the agreement. If Holmes had opened the box under our very noses," Arthur Palmerston continued, "and shown us the thirty-one pieces lying collectively in the box and on the table, the identity of the missing piece would have been established beyond any reasonable doubt. Surely, any natural philosopher present would have qualified this as a 'proof.'"

Well, here indeed was an amusing puzzle! The conversation then got deeper into semantics and the philosophy of inductive evidence. Holmes, meanwhile, had been looking with increasing interest at the Palmerstons' suspended game. Presently he took out of his pocket a notebook, tore off a piece of paper, wrote something on one side, folded it with the writing inside, then wrote something on one of the half-sides, folded it again with the new writing inside, and placed the folded note on one side of the chess table—not on the squares, but on the margin. This was the position:

Black-10

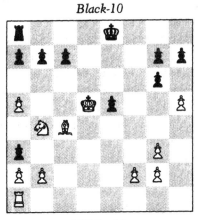

White-11

"You know," said Holmes, "I would like to see this game concluded. It possesses some interesting features."

"By all means," said Sir Reginald. "Why don't you continue it? I would be delighted to watch!"

The two brothers went back to their original seats at the chess table. It was about a minute before either moved. Then Robert Palmerston put his left hand on the king and his right hand on the rook and was about to castle. Arthur raised his head and was about to say something when Holmes sprang up like a tiger and said, "No, no, Mr. Palmerston! Before you move, will you please half-unfold this slip of paper and read aloud what I have written?"

Robert picked it up and read aloud, " 'You can't castle! You really can't, you know!' "

I don't think I have ever in my life seen a man look more surprised! "Good God, Holmes," he exclaimed, "this is really too much! This is indeed a double mystery!"

"Well," said Holmes, laughing, "if you will completely unfold it and read the rest, I think one of the two mysteries will be cleared up."

Robert Palmerston did so, and read, " 'While I have been studying this intriguing position, Robert Palmerston has been studying *me*. I think he knows I am on the trail! I predict that when he goes back to the game, he will pretend to castle, just to test me!' "

"Capital joke!" roared Sir Reginald. "Really capital! And now," he added more seriously, "would you be so good as to explain to us the other mystery? How did you know he can't castle? Were your methods deductive or inductive?"

"Oh, purely deductive," laughed Holmes. "Except for the fact that I did see White's last move as I entered the library. Were you aware of that, Mr. Palmerston?"

"Why, certainly," replied the younger Palmerston. "If you hadn't known White's last move, then you couldn't have known that I can't castle."

"Correct," replied Holmes. "Only how did you know that?" Holmes inquired as he turned to Robert Palmerston.

"Oh, Holmes," was the answer, "this last month I, too, have learned a little about retrograde analysis."

"Splendid!" replied Holmes. "Really splendid! In that case, then, you probably already know my analysis?"

"I believe so," said Robert, "only I would like to hear it from your own lips to see how it tallies with mine."

"Splendid," replied Holmes once more. "Now then, we first observe that the White pawns have clearly captured all six missing Black pieces—the one on a5 has captured two, the one on g3 has captured one, and the one on h5 has captured three. Now, White's last move was *not* with a pawn (since it was with a knight), hence it did *not* involve a capture. Therefore, immediately before White's last move, there were no other Black pieces on the board."

"Clear enough," I remarked.

"So then," continued Holmes, "what was Black's last move? If it was with the king or rook, then of course Black can't castle. If neither the king nor the rook moved last, then the last move was made by one of three pawns on a3, e5, and g6. Now, e5 did not move last."

"Why not?" inquired Sir Reginald.

"For the following reasons: The Black pawn on a3 has made at least three captures, if it came from d7; it has made four, if from e7. The pawn on g6 has made one capture. This accounts for at least four of White's five missing pieces. Thus the pawn on e5 did not make *two* captures."

"I'm with you, so far," said Sir Reginald.

"Well then, this means that e5 did not just come from f6, or else it would previously have had to come from e7, making two captures."

"Right," I said.

"On the other hand, it could not have just come from d6, or else either it came previously from e7—again making two captures—or else it came from d7, in which case a3 must have come from e7 rather than d7, and this would collectively involve six captures, including the capture on g6, which is one too many."

"Good," said Robert Palmerston.

"Therefore," continued Holmes, "if e5 moved last, it must have been from e6 or e7. It couldn't have come from e6 or it would have been checking the White king. And it couldn't have come from e7 as then the Black bishop from f8 could never have got out on the board to be captured by a White pawn."

"Capital!" said Sir Reginald.

"Now then," continued Holmes, "we know that e5 did not move last. Hence—still assuming that neither the Black king nor rook moved last—the last move was made with one of the pawns on g6 or a3. And now I will prove to you that in *either* case, the Black king must have moved sometime earlier in the game—though in each case for quite a different reason! Well, suppose g6 moved last from f7. Then the Black king must have moved sometime earlier to let the Black king's rook out onto the board to be captured by a White pawn."

"Clever," I remarked.

"That is the easy case, Watson! Well now, suppose a3 moved last. It must have been from a4. Now comes the remarkable part! Since a3 was just at a4, it must have come ultimately from d7, making its three captures on c6, b5, and a4, *which are all white squares*. And the pawn on g6 has captured on a white square. Thus four of the five missing White pieces have been captured on white squares. Now, the White queen's bishop originally from c1 fell on a black square, so was not captured by the pawn either on a3 or g6. Therefore the four pieces captured by those two pawns include the White pawn originally from d2. This raises a little problem! For this pawn to get captured by the pawn on a3 or the pawn on g6, it would have to have left the d-file, but how could it, since all six missing Black pieces have been captured by a5, g3, and h5? The only possibility is that the pawn from d2 has promoted! So, if the pawn on a3 moved last, then the pawn from d2 promoted. It must have come straight down the d-file, and when it came to d7, the Black king had to move out of check—unless, of course, it had already moved away. Therefore, again Black can't castle.

"To summarize, gentlemen, either the Black king or Black rook has moved last, in which case Black can't castle, or g3 has moved last, in which case the Black king has previously moved to let out the other Black rook, or a3 has moved last, in which case the Black king has previously moved because of the promoting pawn from d2. Which of the three possibilities is the actual one cannot be analyzed—only the players themselves can know that. But in none of these instances can Black castle."

TWO BAGATELLES

At this magnificent demonstration, Robert Palmerston could not refrain from bursting into applause, in which we all soon joined. Holmes, meanwhile, smiled and bowed like a mischievous schoolboy.

Holmes was the first to speak. "You know," he said, "this was one of the finest 'can't castle' positions I have ever come across. Most of the retrograde problems in the literature are of this variety. Indeed, the first retrograde problem I ever solved was a 'can't castle.' "

"Do you remember what it was?" inquired Sir Reginald with interest.

"Oh yes," replied Holmes, "only I think it is too simple to interest you—a mere bagatelle, you know."

"Why don't you show it to us anyhow? It might be amusing to learn what got you started on these problems, and to test ourselves against your beginning skill."

"Very well. Let's see now, I don't wish to disturb this game. Sir Reginald, you wouldn't by any chance have a second set about?"

"Oh, yes, indeed," replied Sir Reginald, as he walked towards a locked cupboard. "Since last month's mishap, when the children mislaid the bishop, we wouldn't think of taking another chance of inconveniencing our guests, you know!"

At this, Sir Reginald took out a second set and a portable board, which he placed on a nearby table. Holmes set up the following position:

Black-11

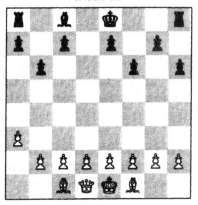

White-12

"It is Black's move," said Holmes. "Can Black castle?"

Since Holmes had described this problem as "simple," I thought I might have a chance of solving it, and so exerted myself to the utmost to do so. And, I cannot help but be proud to say, I was the first to get it. I made a couple of mistakes in exposition, but the mistakes were more in the nature of omissions than of faulty reasoning, hence were not serious. Here is the analysis I gave (with all the gaps filled).

White's last move was clearly with the pawn. Black's last move must have been to capture the White piece which moved before that. This piece would have to have been a knight, since the rooks could not have got out onto the board. Obviously none of the Black pawns captured the knight, and the Black queen's rook couldn't have captured the knight, because there is no square that the knight could have moved from to get to that position. Likewise the bishop couldn't have captured it, since the only square the knight could have come from is d6, where it would have been checking the king. Hence either the king or the king's rook has made the capture. So, Black can't castle.

The company was quite delighted with my solution. Holmes beamed with evident pride at his slow but eager pupil. "You see," said Holmes, "what progress results from application?"

"Show us another," begged Sir Reginald.

"Yes," added Arthur Palmerston, "another simple one. Not perhaps as simple as the last one, but then again, less complicated than the analysis of the game I have been playing with Robert.'

Holmes thought for a while. "I have one," he said at last, "which I think might fill the bill. It is simple, but rather elegant." He arranged the board as follows:

Black-4

White-6

"In this position, it is given that neither White nor Black has captured on his last move. It is Black's move. Can he castle?"

Robert Palmerston was the first to finish this problem. He gave the following neat solution:

"White's last move was not with the pawn on f3, because this would involve a capture. It was not with the rook from e1, since it would have checked the king. Suppose the White king was moved last. It did not make a capture, hence the Black rook which checked it must have just moved to give it check, so Black can't castle. Thus, if White's last move was with the king, then Black can't castle. Now there is, of course, the possibility that White's last move was to castle. Well, suppose this were the case. Then what was Black's last move? If it was with the king or rook, then of course Black can't castle. It could not have been with the bishop, since then White would have had no move immediately before that. Suppose it were with the Black pawn. Then White's preceding move must have been with the pawn from e2, capturing a piece on f3. This means that the bishop on d1 must be a *promoted* bishop! Then the promoting pawn must have come from d7, have passed d2, checked the White king, and made it move! This is contrary to the given that White has just castled.

"In summary: If White's last move was with the king, then the Black rook moved to check him and Black can't castle. If White's last move was to have castled, then Black's last move was with the king or rook, and Black still can't castle."

♟ ♟ ♟ ♟ ♟ ♟ ♟ ♟

SIR REGINALD'S JEST

"As to poetry, you know," said Humpty Dumpty, stretching out one of his great hands, "I can repeat poetry as well as other folk, if it comes to that—"

"Oh, it needn't come to that!" Alice hastily said, hoping to keep him from beginning.

"The piece I'm going to repeat," he went on without noticing her remark, "was written entirely for your amusement."

Alice felt that in that case she really ought *to listen to it; so she sat down and said "Thank you" rather sadly.*

LEWIS CARROLL

We were all delighted with Robert's solution. "And now," said Sir Reginald with a *very* mischievous expression, "I have a little problem for you, Mr. Holmes, which I've been waiting all evening to show you."

Holmes was visibly tired, and did not respond with much enthusiasm.

"The piece I am about to present," continued Sir Reginald, more lively than ever, "was composed entirely in your honour."

Holmes (as he later confided to me) felt that under those circumstances he really *ought* to see it. "Thank you," he said rather sadly.

"Yes, yes, Holmes," continued Sir Reginald, growing more mischievous and enthusiastic by the minute, "this problem is really *your* type of problem. As a matter of fact," he continued proudly, "I composed it myself!"

"Oh?" replied Holmes.

"Ah, yes, indeed!" replied Sir Reginald, gleefully rubbing his hands. "It was inspired by your brilliant solution last month of the missing bishop. My problem, like yours, involves identifying a missing piece. It might be said to be a *variant* of the other."

"Really, now?" said Holmes, with genuinely growing interest. "In that case I really *would* like to see it."

"I thought you would, Holmes, I thought you would," Sir Reginald said as he set up the following position:

Black-9 or 10

White-8 or 9

On a5 Sir Reginald placed a shilling. "The problem is, what is the piece on a5?"

We all looked at the position. Almost at once, I saw the joke and could hardly refrain from bursting out laughing. Arthur and Robert Palmerston caught on a few seconds later, and the four of us found it all we could do to show no visible signs of amusement. But Holmes was quite seriously absorbed in studying the position. He muttered, half to us and half to himself, "Let's see now; Black is in check. What could have been White's last move? Obviously a rook from b7 capturing a Black piece on a7. I guess the next question now is, what was the Black piece? If it was a rook, then Black has promoted earlier . . ."

At this point, we could no longer contain ourselves, and all burst out in a roar.

"I really don't see what's so infernally funny; I really don't," said Holmes, visibly irritated.

"Come, come, Holmes," said Sir Reginald, "I really can't bear to tease you any longer. The missing piece, Mr. Holmes, is obviously the White king."

Well, it did not take long for Holmes—good sport that he is—to join heartily in the mirth. "Capital," roared Holmes, "capital, indeed! I have really been given a good dose of my own medicine. How many times have I not told Dr. Watson: 'In looking for the subtle, be careful not to overlook the obvious.' "

A RETURN VISIT

A week later we had a return visit from Sir Reginald, accompanied by the Palmerston brothers. The visit was not unexpected; Holmes had already set up a chess position in anticipation of it. We were sitting by the fire awaiting their arrival, and Holmes was grinning from ear to ear.

"Why so like the proverbial Cheshire cat?" I inquired.

"Oh, Watson," replied Holmes, breaking into a laugh, "I have a little return jest for Sir Reginald, and I just cannot wait to see his reaction!"

We did not, indeed, have long to wait; almost at that very moment Mrs. Hudson showed in our three visitors.

"Well, well, Holmes, what is this?" inquired Sir Reginald, advancing towards the chessboard Holmes had set up.

"Why, Sir Reginald, this is a little problem composed entirely in *your* honour! The problem is for White to play and mate in one move."

"Remarkable," said Sir Reginald, before he studied the position seriously. "I had no idea that a one-mover could present any challenge!"

Then Sir Reginald turned all his concentration on the problem. After a while he shook his head and said, "I am afraid you've beaten me, Holmes. I really can't get it! The only way White can even *check* Black is with the knight on g4 either moving to h6 or capturing the knight on f6. But neither is a mate!"

"Are you sure of that?" asked Holmes.

"Why of course!" replied Sir Reginald. "In the first case Black can move to h8; in the second, Black can take back the knight with a pawn."

"No, he can't," laughed Holmes, "because Black is going in the other direction! If your side were really White—as it appears—then how could the White king and queen ever have changed places?"*

"*Touché!*" said Sir Reginald with a laugh. "You really fooled me with that one! . . . And now, Mr. Holmes," he went on, "I hope you will give me a chance to restore my honour! This little problem gives me an idea which, I think, might stump *you.*"

"Excellent," said Holmes, who was clearly in the mood for a challenge.

"One thing, though," continued Sir Reginald. "The idea occurred to me only just now, and I have not yet had time to work out the details. I will have to do this experimentally at the board, but I'm afraid that if you see me moving the pieces about, it may give you too good a clue."

"In that case," replied Holmes, "I suggest that the other gentlemen and I retire to the opposite side of the room for a bit, and let you have the board quite to yourself."

"Excellent," replied Sir Reginald, "but no peeking, you know!"

---

* The idea for this problem came from a similar puzzle by Sam Loyd in *The American Puzzlist.*

"No peeking," promised Holmes with a laugh, as we walked across the room.

We sat and chatted pleasantly, and Holmes, true to his word, never glanced in Sir Reginald's direction. I, however, had made no such promise, and I surreptitiously looked over from time to time. But these peeks, I'm afraid, gave me no clues; perhaps I peeked at the wrong times!

About ten minutes later, Sir Reginald called out, "All right, Mr. Holmes, I have it! Pray be seated. The problem again is for White to mate in one."

Holmes studied the position. After a while he said, "I'm afraid you've beaten me this time, Sir Reginald! I really can't see the trick! Had you not shown me this problem immediately after the last, I would have guessed it was again a matter of direction. But even if it were a matter of direction, what help would it be? Regardless of the direction, there is no mate in one!"

"Wrong!" said Sir Reginald triumphantly.

"Wrong?" asked Holmes. "Then would you please tell me what is the mate if my side is White, and what is the mate if your side is White?"

"Neither side happens to be White," laughed Sir Reginald, "as should be evident by the fact that as the board is now oriented towards us, the lower-right-hand corner is a *black* square, instead of a white square as it should be. If you set it right, you will see that there is indeed a mate in either of the other two directions. With White at the right, the pawn on what is now b6 can take the rook on a5 and promote to a bishop or a queen, thus mating the Black king. With White at the left, the pawn on what is now c2 has simply to move to d2 to effect the same thing."

"Rascally!" said Holmes, truly amused. "So that's the real reason you wanted the board to yourself—so you could secretly rotate it ninety degrees!"

"Precisely," replied Sir Reginald.

"Well, Sir Reginald, when it comes to chess jokes, I acknowledge you the undisputed master."

The remainder of the evening we spent studying two serious retrograde problems, from which I learned a great deal—as will the reader, if he studies these analyses with care. Both problems were presented by Holmes.

"Here," said Holmes, "is a position in which it can be proved that White can't castle. The proof is rather simple, but I believe the *reason* why White can't castle will surprise you!"

Black-13

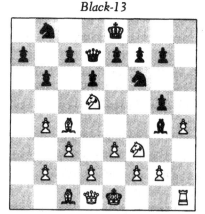

White-15

The reason did indeed surprise us! White is missing only a rook; Black is missing two rooks and a bishop, which was captured on its own square, f8. Therefore the pawn on b4 captured a Black rook and the pawn on g5 captured a White rook. Black must have captured first, since prior to the capture neither of the Black rooks could have got out on the board to be captured by the White pawn. How then did the missing White rook get out on the board to be captured by the Black pawn prior to the White pawn on b4 capturing? The only possible answer is that the rook on h1 must really be the queen's rook! The sequence was this: First the king's rook got out and was captured by the Black pawn, letting out a Black rook to be captured by the White pawn. Then the rook from a1 came round to h1. So the rook on h1 is really from a1! Thus of course White cannot castle.

"That's a pretty problem," said Arthur Palmerston. "I wonder, if the White bishop were removed from c1, would that affect the answer?"

"Let us see now," replied Holmes. "That's a nice question, Palmerston! It would make no difference as regards the final outcome, but the proof would be a bit different. In this case, the rook on h1 could be the king's rook, but if it were, then the queen's rook would have had to get out via the king's rook square, so the king (as well as the king's rook) must have previously moved to let the queen's rook by."

"The next position," continued Holmes, "illustrates a still stranger reason why castling is sometimes impossible."

"It is given that neither queen has ever moved off her own color," said Holmes. "The problem, now, is in three parts:

a—Which side, if either, can castle?

b—If the rook on g1 is removed, would that affect the answer?

c—If the rook is replaced on h1, then what would the answer be?"

Black-13

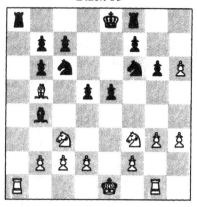

White-13

Here is the analysis that Holmes provided us when we had all given up:

In (a), the piece captured on b6 was not the White queen (who never moved off her own color) nor the bishop from c1 (which never escaped) nor the pawn from a2, because since all three missing Black pieces must have been captured by the pawn on h6, it had no capture to make to get onto the b-file. Therefore the pawn from a2 has promoted. It had no pieces to capture, hence it promoted on a8. Hence the rook on a8 has moved, and Black cannot castle.

Now, the capture on b6 clearly occurred *prior* to the promotion (or the White pawn could not have gone by). Therefore it was not the promoted piece which was captured on b6. That means the promoted White piece is now on the board (since the queen, captured on her own color, the bishop, captured on c1, and some other *original* piece captured on b6 account for the three missing White pieces). What is the promoted piece? Not the bishop on b5, because it could never have escaped from a8 on account of the pawn on b7. Likewise it cannot be a knight, because the pawns on b6 and c7 (the former, we recall, was there *before* the promotion) would have prevented it. Therefore the promoted piece is a rook. If it is the one on g1, then the king must have moved to let it in (the pawns on g3 and h3 couldn't have cross-captured, because all missing Black pieces were captured by the pawn on h6). On the other hand, if the promoted rook is on a1, then again White can't castle. Thus White cannot castle either.

In (b), if the rook from g1 is removed, then there is no evidence that Black cannot castle; the promotion could be avoided by the White king's rook having been captured on b6, but then the king has moved to let it out. (Alternatively, the queen's rook might have been captured, and the rook on a1 could really be the king's rook.) So Black perhaps can castle, but White definitely cannot.

As for (c), in this case Black cannot castle for the same reasons as in (a), but it is possible that the promoted rook is the one on a1, in which case White can castle on the king's side.

To summarize: (a) Neither side can castle. (b) White can't castle; Black may be able to. (c) Black can't castle; White may be able to, but only on the king's side.

MYCROFT'S
PROBLEM

My favorite 'can't castle,' " said Holmes to me one evening, "is the following, which was composed by my brother Mycroft." He set up the board:

Black-14

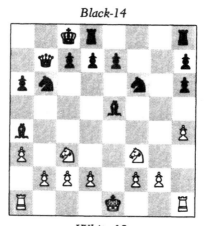

White-12

"In this problem, it is given that White gave Black odds of a queen, and that both White knights are original. White can castle. Can he castle on either side?"

I started to study this position, and Holmes continued, "The interesting thing about this problem is that there is no way of knowing on *which* side White can castle; all that can be shown is that he cannot castle on both sides."

Here is the analysis Holmes gave after I had struggled with the solution for some time unsuccessfully.

58

The piece captured on h6 is not the White queen (which was given as odds), nor the bishop from c1 (which never escaped), nor the bishop from f1 (which is on the wrong color), nor the missing pawn from e2 (which couldn't have made three captures to get to the h-file). Therefore the pawn from e2 has promoted. This promoted pawn is either now on the board or else was captured on h6.

Suppose it is now on the board. Then it must be one of the rooks (since the White knights are both original). Since the King has not moved, the promoted rook could never get to a1, and hence must be on h1. So in this case, White can castle only on the queen's side.

On the other hand, suppose the promoted piece was captured on h6. Then the promotion occurred *prior* to the capture, which means that when the pawn from e2 reached the eighth square, the pawn on h6 was still on g7. Now the White pawn from e2 either promoted on f8, having made one capture, or on e8 or g8, having made two. If the former, then the bishop from f8 was previously captured on its own square (since the pawn from g7 hadn't yet moved), hence the bishop on e5 is promoted. If the latter, then the pawn from b7 must also have promoted in order to become one of the *two* pieces the White pawn captured. To have promoted, the pawn from b7 must have captured a piece on a2 and then either promoted on a1 or captured another piece on b1. Could it have captured a piece on b1? No, because the promoted White pawn was captured on h6 and the White queen was never in the game—which leaves only the two White bishops. They could not have *both* captured on a2 and b1, because both are white squares! So the Black pawn promoted on a1 (which incidentally shows that the White pawn did not promote on f8, because the bishop on e5 would have to be promoted, and it could never have escaped from a1). Since a Black pawn promoted on a1, the rook on a1 has moved, hence White can castle only on the king's side.

In summary: If an original piece was captured on h6, then

the rook on h1 is promoted, and White can castle only on the queen's side. If a promoted piece was captured on h6, then a Black pawn has promoted on a1, hence White can castle only on the king's side.

A LITTLE QUESTION
OF LOCATION

Another stroll to the chess club?" Holmes asked me some days later.

"Why, certainly," I replied, delighted with the possibility of another little adventure.

When we arrived, two gentlemen unknown to us were playing a game. The position was as follows:

Black-9

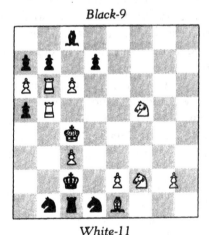

White-11

A White pawn had been carelessly placed on the border between g2 and h2. I was about to inquire on which of the two squares it was supposed to stand when Holmes held up his hand to me. I realized that perhaps he could *deduce* the answer and would like to surprise the players by demonstrating this. So I remained silent and watched his expression with eager anticipation. But Holmes said nothing, and once or twice shook his head a bit hopelessly.

Just then, White placed his hand on a piece and was about to move. "Just a moment, please," said Holmes eagerly. "Would you be so good as to tell me whether this game has been a normal game?"

"A *normal* game?" asked White, astonished. "Just what, sir, do you mean by a *normal* game?"

"Oh," replied Holmes, "by that I simply mean a game in which no pawn underpromotes, a game in which a pawn promotes only to a queen."

"Why, yes," replied White, "this game has been what you call 'normal.' Indeed, so far there have been no promotions at all."

"Ah! Then please permit me," said Holmes, as he leaned over and moved the White pawn to the correct square.

"Thank you," replied White, as he was about to resume play. But suddenly he looked up at Holmes in amazement. "Why, how did you know, sir? How did you know where the pawn lay?"

"Because you were kind enough to tell me," laughed Holmes, who was evidently enjoying this little mystification.

"I *told* you?" replied White, more astonished than ever.

"Why, yes," replied Holmes. "Not explicitly, to be sure, but implicitly. You told me not directly, but by implication."

This remark did not appear to enlighten the players overmuch. Holmes continued, "I soon saw that from the position alone, it was not possible to place the pawn correctly; there were three things I had to know. First, I did not know *for sure* which side was White—though, of course, I could make a pretty good guess. Second, I did not know whose move it was. Third, I did not know whether there had been any underpromotions. Well, when I saw one of you place his hand on a White piece to move it, then of course I knew which side was White and that it was White's move. As to the matter of underpromotions, you yourself were kind enough to tell me. So where is the mystery?"

"But," said Black, "what has all this to do with the location of the pawn?"

"Oh," said Holmes casually, "that was the elementary part, you know. I reasoned as follows:

"Black has just moved; what was his last move? Obviously with the king or one of the knights. It could not have been with either knight, for neither one could have moved from a square which would not be checking the White king. Therefore it was the Black king. It obviously did not come from b3 or d3, nor did it come from d2, since it would then have been in imaginary check from the White bishop. Therefore it came from b2, moving out of check from the White rook. How did White deliver this check? He could not have moved his king away from b4, for he would have been in imaginary check from the pawn on a5. Nor, of course, could he have moved his king from b3. Could it be that the rook itself moved last from c5, d5, or e5 and gave the check directly? Well, considering the other White rook on b6, only if the checking rook *captured* a Black piece on b5. What Black piece could this be? Not a knight, since there are two Black knights on the board, and there have been no underpromotions. Also, not a bishop, because the Black bishop moving on white squares is still on c8. Also not a pawn, because the pawn on a5 comes ultimately from c7, hence no pawn from e7, f7, g7, or h7 could possibly get to b5. Also not a rook, because the rook from a8 could not have gotten out onto the board, but must have been captured on a8 or b8, since neither the bishop on c8 nor the pawns on a7, b7, and d7 have yet moved. The last possibility to examine is the Black queen. This, too, is not possible, because it would have been checking White and could not have come from any square which would not be checking White—even assuming that the rook, in capturing her, had come as far as from e5.

"Thus White's last move was *not* with the rook on b5. Therefore Black's last move was with the king from b2, *capturing* the White piece on c2, which had just discovered check from the rook. This White piece could only be a knight from b4 or a bishop from b3. The former is out, since there are two White knights on the board, and there haven't been any underpromotions. Therefore Black's last move was with his king on b2, capturing a White bishop on c2. The home square of this bishop is f1, hence the pawn between g2 and h2 must really be on h2, otherwise the bishop couldn't have left its home square. That, gentlemen, proves that the pawn is on h2."

"An admirable chain of reasoning," remarked White.

"Did you say," said Black, "that this was the *elementary* part?"

"Well," replied Holmes, in a sort of mischievous half-apology, "perhaps I should have said *fairly* elementary."

♟

A moment later Holmes continued, "Actually, gentlemen, this position is almost identical with one I came across some time ago." Using a board and chessmen from a nearby vacant table, he set up the following:

Black-9

White-11

64

"The conditions are the same as before—that is to say, it is White's move and there have been no underpromotions. The position differs from the preceding in that the pawn which was previously placed ambiguously now stands definitely on g2, whereas the pawn which previously stood definitely on c6 is now placed ambiguously between c6 and d6. The problem is: Does this pawn stand on c6 or d6?"

"Why, on d6, of course," replied one of the two players instantly. His name was Fergusson, and we later found him to be an excellent logician.

"Why?" inquired Holmes

"Because," replied Fergusson, "if it were on c6, we would have the same impossible situation as before."

"Capital," replied Holmes, "but I would like a little more information than that. What I really wish to know is, how does the pawn being on d6 rather than c6 relieve the impossibility?"

"I'm afraid," replied Fergusson, "that your question is not sufficiently precise to admit of a precise answer."

"True, true," replied Holmes. "Well then, let me put the question to you this way: Exactly what was the last move White made?"

"That *is* a precise question," replied Fergusson. "Let me see now. Ah yes, very pretty! What happened was this: The same argument as before establishes that if the Black king just captured a White piece on c2, this piece could only be a bishop. But it can't be a bishop, because of the pawns on e2 and g2. Therefore Black's last move was with the king from b2 to c2, but it did *not* capture a piece. Therefore White's last move must have been with the rook on b5 moving horizontally and capturing a Black piece on that square. What was the Black piece? By the same argument as before, it couldn't be a pawn, knight, bishop, or rook. But a pawn on d6 rather than c6 invalidates the preceding argument that it couldn't be a queen. It could have—indeed it must have—been a queen, which gave check to the White king by coming from c6. But this is possible only if the White rook now on b5 had been standing on c5. Therefore, White's last move was with the rook on b5, moving from c5 and capturing a queen on b5."*

"Excellent reasoning," said Holmes. "I do hope, Mr. Fergusson, we shall have the pleasure of seeing you again."

♟

Holmes and I were just about to leave the club when he stopped at one of the many deserted chess tables on which there was an unfinished game. "Halloa," he said, "this may be interesting. Mr. Fergusson and Mr. Fenton," he called, "if you wouldn't mind stepping over here a moment, I think we may have another intriguing position to analyze. Again we have an ambiguously placed pawn, but this time vertically rather than horizontally." We studied the position:

Black-13

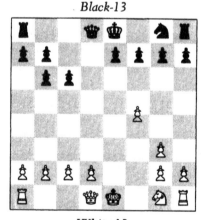

White-13

---

* If the reader finds this explanation difficult to follow, I suggest that he put the Black king on b2, the White rook now on b5 on c5, the ambiguous pawn on d6, and a Black queen on c6—this is how the position was three moves ago. Then the following sequence takes us to the present position: (1) Black queen to b5, check; (2) rook on c5 takes queen; (3) king to c2.

"I wonder," Holmes continued, "if we can deduce on which square it stands? I think we can safely assume that the White is as indicated."

At first, I must say, it looked quite hopeless to me! But the four of us carefully studied the situation together, and gradually the following solution came to light.

Both White's queen's bishop and Black's king's bishop were captured on their home squares. Hence the captures on b6 and c6 were of White's king's bishop and knight— clearly the bishop was captured on c6. The captures by the White pawn on g3 and the one on the f-file (f4 or f5) were of Black's queen's bishop and knight; obviously the pawn on the f-file captured the bishop. Now, Black's queen's bishop did not get out onto the board until *after* the capture on c6. So White's king's bishop was captured before Black's queen's bishop. The pawn on f-file comes from e2; it moved before White's king's bishop got out and was captured. It couldn't have captured on f3, because Black's queen's bishop hadn't got out yet to be captured. Hence it moved from e2 to e4 and later captured Black's queen's bishop on f5.

The sequence was this: Pawn from e2 moved to e4, then White's king's bishop got out to be captured on c6, then Black's queen's bishop was captured on f5. So the White pawn must be on f5.

"TO KNOW
THE PAST"

It is a curious thing," remarked Holmes to me a few evenings later, "that to know the past, one must sometimes first know the future."

"Oh!" I replied, trying to guess the meaning of this typically enigmatic remark. "Can you be a bit more specific, Holmes?"

"Why, yes," he replied. "I have a very specific incident in mind, as a matter of fact. It occurred to me as a result of two recent events. One was our little adventure of the ambiguously placed pawn. The other was the fact that one of my cases took me this morning into a research laboratory. In the director's room was a sign which caught my eye:

TO KNOW THE PAST, ONE MUST FIRST KNOW
THE FUTURE

"These two incidents recalled the following: About seven months ago, I was working on a very serious case, the clue to which I knew was to be found in one of the many rooms in Lord Bottomley's house. I called one evening at the estate, but found, to my disappointment, a party in progress. However, Lord Bottomley, who himself had a considerable interest in seeing this case resolved, was most pleasant and cooperative. He introduced me as a guest, and later told me privately that I should feel perfectly free to pursue my investigations in any part of the house I desired.

"Well, Watson, after looking through several of the

empty rooms, I finally came upon the clue I was after. The problem was now solved. Solved, that is, pending a certain message I was expecting from Detective Lestrade. I had nothing left to do but to wait for the message, which I was virtually certain would corroborate my own findings.

"I was not in a particularly sociable mood that evening and did not care to rejoin the guests. To pass the time, I slowly wandered through the rooms of a rather deserted wing of this very fascinating house. I came to one rather small room—a sort of library-study. In the center was a possibly unfinished chess game. Two inadequately extinguished cigars indicated that the players had but recently left.

"To my great surprise, Watson, a pawn was placed not on the *border* between two squares but on a *corner* between four squares—c4, c5, d4, and d5! The position was this:

Black-2

White-4

" 'Tut, tut,' I said to myself, 'how careless can one be?' Anyway, I thought I would see if I could deduce on which of the four squares the pawn lay, and then have a little fun by leaving a note for the players—if they should return—telling them where the pawn should be. But I soon saw that, as matters stood, the problem was unsolvable. I had ample evidence, with which I will not bore you, to establish which side was White, but I still couldn't know where the pawn should be unless I knew something about the *future* of the game!"

I was more mystified than ever. Holmes continued, "At that very moment I heard footsteps behind me, and a voice said, 'A most delightful house indeed, but now let us resume our game,' upon which two gentlemen entered the room, nodded to me, and walked towards the chess table. I was delighted to find some hope of resolving this mystery, when suddenly my eye caught the butler in the corridor making his way towards me. I quickly stepped outside to meet him, and he said, 'Sir, I have been looking for you everywhere. There is an urgent message for you outside in the garden.' I briskly sprinted out to the lawn, and there was a boy who handed me my awaited message from Lestrade. It indeed confirmed my own findings, and my mind was completely at rest on that issue. There was no emergency involved, so I had the rest of the evening to myself.

"I went back to the room to clear up the still-unsolved chess mystery. When I arrived, I found, to my great disappointment, that the game had been concluded; the players were now involved in the beginning stages of a new game. All traces obliterated! But then I had an idea: 'Gentlemen,' I asked, 'would you mind telling me who won the last game?'

"'I did,' replied one of them. This, Watson, solved the problem! The mystery is really exceedingly elementary, and I think you might enjoy trying your own hand at it."

I stood awhile in thought studying the position which Holmes had reproduced for me. "Holmes," I said suddenly, "I believe that you have inadvertently or deliberately withheld a vital piece of evidence from me."

"I most definitely did not," retorted Holmes decisively.

"My dear Holmes," I replied, "do you realize that you have not even told me whether it was White or Black who won the game?"

"Of course not," replied Holmes. "That happens to be most irrelevant. As a matter of fact, I myself did not know whether it was White or Black, nor did I need to. I would imagine it was White, since White appears to have a winning position, but I did not inquire whether it was White or Black who won, but only which of the two players won."

I thought about this, but could not fathom how it could be relevant which *person* won. Finally, I said, "Holmes, I'm afraid I'll have to ask you for a hint. Suppose it had been the other player who told you that he won. Could you have still found the solution?"

"Certainly," said Holmes.

"And would the solution have been the same or different?"

"The same, of course."

Well, here was indeed a puzzler. "Holmes," I finally said, "I'm afraid I'll have to give up."

"Well, well," said Holmes. "Actually, this is the type of problem, I think, which if one cannot solve immediately, one is not likely to solve at all. The whole point, Watson, is this: I asked for more information than I actually needed. It did not matter to me which *side* won, and certainly not which *person* won, only that *somebody* won. In other words, the game did not end in a draw. More specifically, the game did not end in a stalemate. Now, Watson, if it was White's move, then regardless of where the White pawn stood, White would have no possible move to avoid stalemate. Hence it has to be Black's move. But what could Black's move be? The only possibility is that the ambiguously placed pawn really stands on c4 and the Black pawn takes it *en passant*. Therefore the White pawn stands on c4."

"Ah, Holmes, I see it now." Then another thought occurred to me. "But Holmes, does not the validity of your solution presuppose that White's last move was with the pawn on c4—that it had just moved from c2? What evidence did you have that this was White's last move?"

"You have it the wrong way around, Watson! My solution did not *presuppose* this move; it *proved* that this must have been the last move! If this had *not* been the last move, then the position *would* have resulted in stalemate, which it didn't. A simple case of *reductio ad absurdum.*

"So, Watson," continued Holmes with a chuckle, "is it not amusing how it sometimes happens that to know the past, one must first know the future?"

A STUDY IN
IMAGINARY CHECKS

The next time Holmes and I visited a chess club, I learned a valuable lesson—namely, how it is sometimes possible to arrive at a perfectly correct conclusion using faulty reasoning. What happened was this.

When we arrived, the clubroom was deserted. Several chessboards were lying about, some games finished and others not. One particular position caught my eye:

"I wonder if White can castle in *this* position," I laughed. "Not that I can see any reason why he would *want* to, but I wonder if he could if he did?"

Holmes looked at the position for a moment and replied, "That question is quite easy to answer. What do *you* think?"

I looked again. "Ah," I said triumphantly. "White can't castle."

Holmes looked at me and asked, "Why not?"

"For a very simple reason," I laughed. "For White to castle, it must be White's move. And it is not White's move."

"How do you know it is not White's move?" inquired Holmes.

"Because," I laughed, "if it is White's move, then Black has moved last. But every square the Black king could move from involves an imaginary—or impossible—check."

"That is not necessarily true, Watson. You have indeed arrived at the correct conclusion—that White can't castle—but your reasoning is inadequate. I would analyze the situation this way:

"If White can castle, then it is White's move. The only possible way for it to be White's move is if the Black king just came from a3. This is possible, provided White is moving in the opposite direction than he appears, the last move of White having been a pawn from b2 capturing a piece on a1 and promoting to a rook. But in that case, White can't castle anyway, because he is going in the wrong direction. Thus the real reason why White can't castle is that either it is not his move *or* he is going in the wrong direction."

"Of course," I said.

Holmes rearranged the board:

"Actually, Watson, the previous puzzle is the second time we have encountered a situation in which what appears to be an imaginary check may instead be the result of an underpromotion. The first, if you recall, was White's last move in our little mystery of the missing piece. The board I have just set up has two other pseudo-imaginary checks which sometimes fool people."

Holmes continued, "In this position, Black could have come from either of the squares a6 and c6. A knight promotion could account for either. If Black came from c6, then a White pawn from b7 advanced to b8 and became a knight. If Black came from a6, then a White pawn from a7 captured a piece on b8 and became a knight."

AN UNSOLVED
PROBLEM

Watson, tomorrow I must leave for the Continent,"
Holmes said unexpectedly to me one evening at Baker
Street. "I am on a case of international importance, and I
have no idea how long I shall be gone. It will probably be
several weeks, and may be some months."

"What will I do for chess adventures?" I asked sadly.

"By now, Watson," said Holmes good-naturedly, "I think
you can fairly well function on your own. Just keep your eyes
open and your mind alert.

"Though before I leave, Watson," he continued, "I wish
to discuss with you a problem which has been on my mind
and which I have been unable to solve."

"You expect *me* to solve it?" I asked incredulously.

"No, Watson, I'm not sure that *anyone* can solve it," he
replied thoughtfully. "Indeed, I'm not sure that it *has* a
well-defined solution. The solution seems to lie somewhere
in the borderland between chess, logic, philosophy, linguistics, semantics, and law."

"Sounds like a most intriguing combination," I replied.
"Please let me see it."

"First I must give you a bit of historical background,"
said Holmes. "Recently I have been doing some research on
the evolution of the game of chess. The rules have changed
many times throughout the centuries. Here, it is the last
change which is relevant."

"What change was that?" I inquired.

"It concerns the rule for pawn promotion. Before the last
change, the rule read: 'When a pawn reaches the eighth

square, it is converted to any piece except a pawn or king.' However, the old rule neglected to say that the pawn must promote to a piece *of the same color.*"

"Why would anyone ever want to promote to a piece of the *opposite* color?" I asked in my typically practical fashion.

"Oh, I don't know, Watson, but that's really not the point. I believe the rules of a game like chess should be absolutely clear and unequivocal. It certainly is not *likely* that one would want to promote to a piece of a different color, but it *may* happen in a rare case that it is to one's advantage to do so. Indeed, this *did* once happen, and that's why the rule was changed. It was during a tournament, and White mated Black by promoting to a Black knight."

"How did that happen?" I asked.

"Oh, it was something like this," said Holmes, and he set up the board:

Black-2

White-3

"There were many more pieces on the board, but for illustration this position will suffice. As the position stands, White cannot win in one move if the game is played according to our present rules of promotion, but before the change, White wins in one move by advancing the pawn and promoting to a Black knight.

"At any rate," continued Holmes, "I really wouldn't have cared if such a case *never* came up in actual play. When I came across this change of rule, my first thought was what a field day it would be for retrograde analysts to compose problems according to the archaic rules of pawn promotion! I hope some future retrograder will do this.

"Then," he went on, "the following curious problem occurred to me. Let us say the following game was played in the old days when it was allowed to promote to a different color:

Black-3

White-3

"Now, supposing it is given that it is Black's move and that Black has never moved his king. My question is: *Can Black castle?*"

I looked at the position, and quickly got the point: White's only possible last move was with a pawn promoting to the Black rook on a8. Now it is given that the Black king has not moved; the crucial question then is whether or not the promoted Black rook can be said to have moved.

"I wonder if the rules for castling are finely enough stated to settle that point," I suggested.

"Perhaps not, Watson, and yet I don't know! The rule for castling seems reasonably explicit: Castling is permitted provided that, one—neither the king nor the rook has moved; two—the king is not in check; three—the king does not pass over any checked square. Conditions two and three are clearly met; the whole problem lies with condition one. Now we are given that the Black king has not moved, so the whole question, as you pointed out, lies with the rook. I would tend to say that the promoted rook *has not yet had time to move*, hence that Black can castle."

"And I would tend to say the opposite," I replied. "I would say that the Black queen's rook was moved off the board when it was captured, and moved back on the board when it was reinstated by a promotion. So I would say that the rook has moved."

"But is it really the same rook?" asked Holmes.

Well, this is indeed a puzzler! I'm sure different readers will take different sides on this issue. Holmes and I spent most of the evening discussing it, but of course we came to no definitive conclusion. We run into this problem a bit more deeply somewhere in Part II. Meanwhile, this is an appropriate place to break my narrative.

PART II

MARSTON'S ISLAND

♟ ♟ ♟ ♟ ♟ ♟ ♟ ♟

ABOARD SHIP

MAY 3, 1895 Here we are, Holmes and I, aboard a luxury liner bound for a tiny island in the East Indies. It came about as follows.

I last left the reader on the night before Holmes's departure for Europe. He was gone for about three and a half months and returned quite unexpectedly last week—April 28. It was a beautiful day, and I was strolling in the park. "I thought I would find you here," came that familiar voice from behind me. "I arrived home about two hours ago, bag and baggage, found you gone, and fancied you would take advantage of a day like this."

"Why, Holmes," I joyfully responded, "I had no idea you were back! Tell me all about your trip!"

"Later," laughed Holmes. "But for now, tell me, how would you like to take a trip to the East Indies?"

"The East Indies!" I replied in astonishment. "Surely you are jesting. Or is it that the case takes you there?"

"Oh no," replied Holmes. "I brought that affair to a happy resolution on the Continent. It turned out to be far less intriguing than I had anticipated—indeed, it proved to be a most routine case. Still, my presence was vital and all the criminals have been apprehended. So, I repeat, how would you like to go on a trip to the East Indies?"

"But the expense!" I protested. "My present bank account would hardly make such a luxury practicable!"

"As a matter of fact, Watson, the trip will cost us nothing, and there may be a nice profit to boot."

"Tell me more," I responded with growing interest.

"I will indeed," replied Holmes, "but what do you say to first a little lunch? I had hardly any breakfast this morning."

About an hour or so later, comfortably settled on an outdoor terrace, after a good lunch and a complete accounting of the recent case, Holmes continued, "Well, Watson, it is all in connection with Colonel Marston. You recall that some time ago he bought his brother's house on a very small island in the East Indies?"

"Yes, I recall that."

"Do you know why he bought it?"

"I presume as a retirement spot."

"That's part of it, Watson, but not quite the whole story. Do you know anything of Marston's ancestry?"

"Not a thing," I replied.

"Well, his great-grandfather, Captain Marston, was a rather famous individual."

"Famous in what capacity?" I inquired.

"Well, to put it bluntly, he was famous as a pirate."

"A *pirate!*" I replied in astonishment. "Did you say *famous* or *infamous?*" I added with a smile.

"I suppose either adjective would be applicable, though I personally prefer to think of him as famous. He was really a most unusual sort!"

"In what way?" I asked, somewhat skeptically.

"Oh, Watson, he was a sort of Robin Hood type—robbing the rich and giving to the poor. Of all the buccaneers in history, I really find him one of the most sympathetic. Now, I certainly do not wish to condone his obvious disrespect for the law. Nevertheless, in all justice to him, it must be added that for an outlaw, he was a remarkably humane one! In the first place, his piracies were all bloodless—indeed, there is no evidence that he ever did physical violence to anyone. Secondly, he never robbed a ship of its entire cargo; he treated all his prisoners with the utmost consideration; and it was an absolute point of honor with him to see that they always got to safety. In fact, he more than once risked his life on this account! Thirdly, he always rescued any ship in distress. As I say, he was as humane as any buccaneer I know.

"I did a considerable amount of research on him some time ago," continued Holmes, "and I really think his life would make a magnificent historical novel. He was a highly cultivated individual with a mania for collecting rare manuscripts, and a passion for chess. It is this latter passion which I think may be relevant for our forthcoming venture."

"*Our* venture, Holmes?" I asked with a smile. "Are you not deciding in my behalf?"

"I really think you will want to go, Watson, when you have heard the rest of the story."

"I am all ears," I confessed.

"Well, then, old Captain Marston gave away almost all his booty to poor deserving families. Some, however, he held in reserve, and we think he buried most of this reserve shortly before his death on the small island in the East Indies on which he had made his home.

"The island," continued Holmes, "has been for years in the possession of Colonel Marston's brother Edward. It has been established almost beyond doubt that the treasure is buried *somewhere* on the island, and the value and contents appear to be fairly well approximated—about two hundred thousand pounds in gold and jewels. However, just where the treasure is buried is totally unknown, and to excavate the whole island would cost considerably more than it is worth. Thus the treasure has lain slumbering for close to a hundred years.

"Now," Holmes went on, "the last time Colonel Marston was in London—the time we had our first chess adventure with him and Sir Reginald—he told me, several days later, that Edward had just got married to an American lady who had absolutely no taste for the tropics—indeed, she wanted to live close to her family in Boston. So Edward sold the entire estate to his brother for a relatively small amount, with the understanding that if the treasure was ever found, it would be equally divided between the two brothers.

"Now I come to the exciting part, Watson! A week ago in Paris I received word from Colonel Marston that there is now some hope of locating the treasure, and he requests our help. It appears that Marston has found amongst the rare volumes in the Captain's old library a map hidden between the pages of one of the manuscripts. I am not clear as to the details, but Marston believes that the map may solve the mystery. It is in the form of some chess diagrams with accompanying messages in some sort of code. Marston said that what is called for is a combination of cryptography and retrograde analysis. So he has invited us to come to his island with all expenses paid, regardless of whether we successfully locate the treasure or not. And if we do, he and his brother promise us a liberal share."

So here we are, first-class passengers, bound for the East Indies! So far the weather has been as fair as can be, and this promises to be a delightful voyage. The ship is just teeming with chess players! Games are in progress all over: on the decks, in the lounge, in the dining room, just about everywhere. So I anticipate a good lot of fun! I will, of course, keep a faithful account of any chess adventures we come across.

By now, the reader who has faithfully followed the adventures of Part I should have no small degree of competence in retrograde analysis. Hence I shall write up the ship's adventures somewhat differently than I did there. More specifically, I shall not include the solutions together with the narrative, but shall merely pose the problems, for the reader to try to solve for himself. I shall, however, give all solutions at the end of the book.

THE MYSTERY OF THE INDIAN CHESS SET

MAY 4. Today we had our first chess adventure, and an unusual one it was!

Aboard ship are two brothers, natives of India. They own a magnificent but curious chess set from their country. There is no difficulty recognizing the shapes of the pieces—the main difference being the rooks, which are in the form of battle elephants—but the colors are unusual. Instead of the usual black and white or red and white, the colors of this set are red and green.

Holmes and I came across the following position:

Green-15

Red-15

The players had temporarily abandoned the game and gone for a stroll on the ship. Several other chess enthusiasts were looking at this unusual set and wondering which of the colors was really White and which Black; some conjectured one and some the other. Holmes looked at the position for a while and announced, "Gentlemen, it turns out to be quite unnecessary to guess about the matter; it is *deducible* which color is really White."

And so I leave the problem for the reader: Which color is really White?

ANOTHER QUESTION
OF LOCATION

MAY 6. **O**ur next adventure, which involved a problem with a very pretty solution, was again a case of a displaced pawn. Holmes and I were strolling on deck and came across two players seated at this position:

Black-9

White-7

We paused to look. Just then Black castled—whereupon Holmes correctly adjusted the pawn. Naturally both players were surprised, and subsequently intrigued when Holmes explained how he knew.

How *did* he know?

HOLMES SETTLES
A DISPUTE

The moment Alice appeared, she was appealed to by all three to settle the question, and they repeated their argument to her, though, as they all spoke at once, she found it very hard to make out exactly what they said.

LEWIS CARROLL

MAY 8. Holmes has already acquired quite a reputation on board as a chess detective! He is much talked about, and our next adventure, a rather humorous one, only added to his fame.

We came across a deserted game, with several players standing about arguing whether Black could or could not castle.

Black-15

White-14

One of the observers argued that Black could castle on the king's side but not on the queen's; another, that Black could castle on the queen's side but not on the king's; a third, that he could not castle at all. Now, each of the three observers had been present at certain stages of the play, but none had been present during the entire game. Thus, each one had remembered different facts which he used in support of his argument.

At Holmes's approach, all three rushed up to him to settle the argument. As all three stated their cases simultaneously, Holmes and I found it a bit difficult to make out exactly what they said. But after the confusion subsided somewhat, Holmes was able to extract the following data concerning the history of the game:

1 White gave Black odds of a rook.

2 White has not yet moved either knight.

3 No promotions have been made.

4 White's last move was pawn from e2 to e4.

Armed with these facts, Holmes studied the position anew. After a while Holmes said to the three disputants, "Gentlemen, all of you are wrong! Assuming you have reported these four facts correctly, it would follow that Black can castle on either side—though it is not his move now. But after White's next move, Black can castle on either side."

"This can be *proved?*" asked one of the three in astonishment.

"Why, yes," replied Holmes.

"It can be proved that Black *can* castle?" the disputant repeated incredulously.

"Certainly."

"This is truly remarkable, Mr. Holmes. I have known many positions in which it can be proved that a given side can't castle, but I have never before come across a situation in which it can be proved that it *can.*"

"Neither have I," replied Holmes.

"What puzzles me," continued the disputant, "is this: I

can see how to prove that a king or rook has moved, but I cannot for the life of me see how it can be proved that it *hasn't* moved."

"In this case the proof is quite elementary," replied Holmes.

Can you see how to prove it?

THE CASE OF
THE DROPPED PAWN

MAY 9. Today Holmes was appealed to to settle another question when we came across the following game in progress:

Black-15

White-14

A White pawn had been accidentally knocked off the board. Neither player could remember for sure on which square it stood. Holmes studied the position awhile. "I'm afraid it is not determinable from the position alone where the pawn must be. I will need to know something about the history of the game. Are there any facts you can tell me?"

"Well," replied one of the players, "neither king has yet moved. Is that any help?"

"Let's see now," replied Holmes as he studied the position further. "Why yes, indeed, it is of help! I now know where the pawn must be."

Where?

FROM WHERE?

MAY 12. Today's adventure was, from a purely chess-theoretical point of view, the most interesting we have yet had on this trip.

We were strolling on deck, turned a corner, and quite suddenly came upon the following game:

Black-14

White-15

White was just removing his hand from the pawn on f4. Therefore this pawn had just been moved. But we did not see from which of the possible squares—f2, f3, or g3—he had moved it.

We sat down to watch the game. Black did not respond for quite a while, and Holmes studied the situation with much attention. Suddenly he asked, "Gentlemen, are there any promoted pieces on the board?"

"Why, no," replied one of the players.

"Ah, that solves the mystery," said Holmes.

"What mystery?" asked the other player.

"Well, you see," replied Holmes, "as I turned the corner, I saw you concluding your move with the White pawn, but I did not know from what square you had moved it. Now I know!"

How did Holmes know?

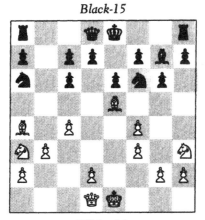

DIFFICULT?

MAY 14. Today Holmes and I came across the following deserted game:

Black-15

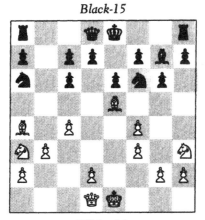

White-14

We looked at the position for a while.

"This looks like a difficult position to arrive at," I remarked.

At this, Holmes burst into an unexpected laugh. "You know, Watson, I could well give a Johnsonian answer to your Boswellian remark. Do you know the story of Johnson and Boswell at the violin concert? Well, the violin virtuoso had finally managed to struggle through an extremely difficult composition. 'That piece must have been very difficult,' said Boswell to Johnson. 'Difficult?' replied Johnson. 'Why, I wish it had been impossible!'

"I might make a similar remark about this chess position. I might indeed say: 'Difficult? I wish it were impossible!' However, it is hardly necessary for me to say this, since in point of fact the position *is* impossible! I have no idea who played this game, but whoever played it certainly does not know the rules of chess."

How did Holmes know this position is impossible?

♟ ♟ ♟ ♟ ♟ ♟ ♟ ♟

THOUGHTS OF
A LOGICIAN

MAY 16. A delightful surprise! The reader may remember the logician Fergusson from Part I, and the adventure of the carelessly placed pawn. Well, he is aboard ship.
Holmes and I spent the entire day chatting with him. He is
a most remarkable individual! He is interested equally in
philosophy and mathematics and is doing significant research in the foundations of mathematics. He is a student
and ardent disciple of Gottlob Frege.

"Since you like logic puzzles," said Holmes to Fergusson,
"have you heard the one about the island inhabited by
knights and knaves? The knights always told the truth and
the knaves always lied. One day a stranger passed a garden
in which stood three inhabitants of the island—let us call
them A, B, and C. The stranger asked A: 'Are you a knight
or a knave?' A mumbled an indistinct reply which the
stranger could not make out. The stranger then asked B,
'What did A say?' B replied, 'He said he was a knave.' C
then said to the stranger, 'Don't believe B, he is lying!' The
problem is to determine whether B is a knight or a knave,
and whether C is a knight or a knave."

"Yes, that is a well-known puzzle," replied Fergusson.
"The solution is—"

"Just a moment," I interrupted. "I have not heard this
problem before. Would you give me a minute to think about
it?"

"Why, surely," replied Fergusson.

I thought about the matter awhile and came to the following solution: "Since C contradicts B, then C and B must

be opposite—i.e., one of them is a knight and the other a knave. Which is which? Well, B said not that A is a knave, but that A *said* he was a knave. Could A really have said that he was a knave? Certainly not; if A was a knight, he would never have lied by saying that he was a knave. And if A was a knave, he would never have made the truthful statement that he was a knave. Therefore A never said that he was a knave! Hence when B said that A said he was a knave, then B was lying. Thus B is a knave and C is a knight."

"That is correct," said Fergusson. "You know," he continued, "there is one feature of this problem I never have liked—namely, that C is sort of irrelevant—irrelevant in the sense that the moment B spoke, it was possible to deduce his nature without knowing what C said. Logically there is nothing wrong with this, but I regard it as an *aesthetic* weakness. Now, I have thought of an improved version of this problem which is free from this undesirable feature. Would you care to hear it?"

"Why, certainly," said Holmes.

"Well then, instead of the stranger asking A whether he is a knight or knave, let his question to A be: How many knights are amongst you? Again, A answers too indistinctly to be understood by the stranger. So the stranger again asks B, 'What did A say?' And B replies, 'A said there is *one* knight amongst us.' And again C says, 'Don't believe B, he is lying!' Again the problem is, what are B and C?"

Holmes and I thought about this for a while, and we agreed that it was indeed an improvement. I think the reader might enjoy trying to solve this problem.

"You know, Holmes," I said a moment later, "I think your controversial problem about castling might well appeal to a logician! Why don't we tell it to Fergusson? I'd be interested in a logician's views on this matter."

And so we told Fergusson about Holmes's unsolved problem—the one I presented to the reader at the end of

Part I. We explained how in days of yore, a pawn was allowed to promote to a piece of a different color, and we set up the position. Holmes repeated his argument that Black could castle, since the promoted rook has not yet had time to move. I repeated mine, that the rook had been moved off the board and then back on again.

Fergusson was vastly amused by the whole problem, and considered Holmes very ingenious for having thought of it. "Actually," said Fergusson, "the problem goes deeper than meets the eye. The real problem, as I see it, is how, exactly, you would define the notion of a *piece*. I guess you, Dr. Watson, would identify the piece with an actual physical object, would you not?"

"Of course," I replied. "What could a chess piece be if *not* a physical object?"

"Ah, that's the whole difference between your point of view and that of Mr. Holmes! Mr. Holmes, I think, is, without realizing it, a Platonist like myself; you tend to be more of a nominalist. To us Platonists, the piece itself is not a physical object; the physical object that you handle is merely a *symbol* for the piece. The piece itself is an idealized mathematical entity."

"I'm afraid that *is* beyond my ken," I admitted. "I never was very good at philosophy."

"But the issue is an important one," replied Fergusson, with growing enthusiasm. "Your nominalistic identification of the piece with its mere material representation can lead to serious problems! For example, suppose that during a game of chess someone removed a White pawn from its square and replaced it by another White pawn from the chess set. Would you say that the pawn had *moved?*"

"Well, no," I confessed. "Under those circumstances I would say not. But," I stubbornly persisted, "that is a different situation from the one at hand. In Holmes's problem, the Black rook has been previously captured and has been off the board for some time, and then reinstated via a promotion. Under these circumstances I'd say the rook *has* moved."

"How do you know it was for *some time?*" responded Fergusson. "For all you know, White's last move may have been with the pawn not from a7 but from b7, capturing a Black piece on a8 and promoting to a Black rook—indeed, the pawn may have captured the very Black rook to which it promoted! Well, suppose that is what has happened—that the pawn promoted to the very Black rook it had just captured, and furthermore that the rook had never moved previously. Under those circumstances, would you still say that Black can't castle?"

Well, we all had a good laugh over that one! Logicians do come out at times with the wildest phantasies. But it is great fun.

Later that day, Fergusson had a problem he was eager to show us. "This problem," he said, "combines logic and chess in an interesting manner. The problem is this: Is it possible to design a position so that it can be proved that White has a mate in two moves, yet that it is impossible to exhibit the mate?"

Both Holmes and I were quite puzzled as to exactly what Fergusson had in mind.

"Could you be a bit more explicit?" I asked.

"Well, yes," replied Fergusson. "I have in mind a concrete position. In this position it can be proved that White can play and mate in two. By this I mean that it can be proved that there *exists* a move for White such that given any reply of Black, White then has a mating move. And yet there exists no first move for White such that it can be proved *of that move* that given any reply of Black, White then has a mate on the next move."

"I wish my brother, Mycroft, were here," said Holmes. "He has more of a flair for abstract reasoning than I. Indeed, this type of problem is just his cup of tea! As for myself, I have had little training in this highly refined technical type of reasoning. However, I must confess that what you describe sounds to me impossible! After all, chess is a finite game; at any stage there are only so many moves that can be made. Hence, it would seem that one can examine finitely all the many possible outcomes of the next two moves. If one of White's possible first moves leads to mate in each of the possible outcomes starting with that move, then this must be a proof that White can mate in two, and moreover must be a proof *for that particular move*. So I am completely puzzled by what you say!"

"Well, now," replied Fergusson, "your argument contains a subtle but deceptive fallacy. However, rather than continue the discussion on this abstract level, let me show you the concrete position I have in mind." He arranged the chessmen as follows:

Black-3

White-8

"I claim that in this position, given that it is White's move, it can be proved that White can play and mate in two, but that one cannot exhibit the mate."

Holmes and I studied the position for several minutes. Suddenly Holmes got it. "By Jove, Mr. Fergusson, you are right, absolutely right! Really brilliant!

"You remember, Watson," Holmes continued, "that

problem I once showed you in connection with the motto 'To know the past, one must first know the future.' Well, I think Fergusson's problem nicely illustrates the more usual principle: To know the future, one must first know the past."

If the reader finds himself baffled at this point, he will become pleasantly unbaffled when he reads the solution.

♟ ♟ ♟ ♟ ♟ ♟ ♟ ♟

A QUESTION
OF PROMOTION

MAY 18 (3:00 P.M.) **B**y now, Holmes and I are pretty well acquainted with virtually every chess player aboard. Today we came across the following game in progress:

Black-14

White-15

We knew both players somewhat. When we sat down, it took quite a while before the next move was made. Suddenly Holmes said, "What an interesting game, gentlemen. I perceive there is a promoted piece on the board."

"Quite true," replied White, "only how on earth did you know?"

"Elementary, Mr. Wilson—really elementary," said Holmes, who then explained the solution.

The solution is indeed rather elementary.

SHADES OF
THE PAST

MAY 18 (3:20 P.M.) The effect that Holmes's solution had on Mr. Robinson, who was playing Black, was rather alarming. He seemed to be in a state of shock. "Are you ill?" I inquired.

"Why, no," replied Mr. Robinson, "it's just that this incident reminds me of an extremely painful experience I once had."

"Would you care to tell us about it?" asked Holmes in his typically ingratiating manner.

"Why, yes," replied Robinson. "It was about three years ago, aboard a liner much like this one. There was one particular passenger aboard who arrested my attention from the start. I never got to know his name; he was extremely aloof, considering the generally informal atmosphere of the cruise. He had one constant companion with whom he came aboard; the two kept almost exclusively to themselves."

"What was there about the passenger which so arrested your attention?" Holmes inquired.

"It was a remarkable combination of characteristics. I could not help but overhear many conversations between the two. The passenger in question discoursed most learnedly on mathematics, astronomy, philosophy, and law—I would judge that he was a professor. But there was a certain indefinable quality about him, a quality I can only describe as 'sinister.' It was a quality I *sensed* rather than knew, but I sensed it to the depths of my being. He had a very furtive manner, and seemed to eye with suspicion just about everyone around him. Indeed, if I were to judge by his manner rather than his conversation, I would have taken him for a criminal rather than a professor."

"Perhaps he was both," said Holmes. "Can you give me a physical description of him?"

"Oh, yes," replied Robinson, "I remember him most vividly! I would describe him as extremely tall and thin with deeply sunken eyes. He was clean-shaven, pale and ascetic-looking. There was something reptilian about him—particularly about the strange way his protruding face slowly oscillated from side to side."

"A most singular description," said Holmes, casting me a meaningful glance. "And now, how did my analysis of this chess game remind you of this individual?"

"Why, it was the situation," replied Robinson. "I was playing a game of chess with a fellow passenger. At about the middle of the game, the two strolled by and stopped to watch. They did not sit down in a friendly fashion, as did you and Dr. Watson; they just stood at a rather respectful distance and watched. At one point the passenger in question said, not to me or the other player, but to his companion, 'Curious game; one of the pieces on the board is promoted.' Although he said this rather softly, and in almost a conspiratorial tone, I could not help but overhear the remark. 'Do you happen to know whether it is Black or White?' I asked him. He looked at me with an icy stare and replied, 'I was not addressing you, sir,' upon which he and his companion brusquely walked away. This was about the last I saw of them; the few times we saw them at a distance, they appeared deliberately to avoid us."

"An interesting episode indeed," said Holmes, "albeit an unpleasant one. But, Mr. Robinson, I am still at a loss to understand why this memory should be *that* painful to you."

"Well, Mr. Holmes, I have not told you the whole story," replied Robinson. "I'm a bit hesitant to speak of the rest."

"I'm sure I have no desire to intrude into your private affairs," said Holmes, rising.

"Oh, Mr. Holmes, please be seated," pleaded Robinson. "It's not that the affair is *private*; it's just that—well, quite frankly, it's just that I feel a bit foolish!"

Holmes sat down again with an interested and genuinely sympathetic expression. "You see," continued Robinson, "other events occurred on this ship, events which I could not but sense might be related to this passenger, but I had not the slightest shred of objective evidence! Just a vague, though extremely strong, feeling. Perhaps I am simply suffering from an over-lively imagination!"

"Just what were these events?" asked Holmes firmly.

"Well, Mr. Holmes, two passengers died aboard ship under very mysterious and suspicious circumstances. Foul play was suspected, and there was a police investigation. However, no really conclusive evidence turned up, and the case was soon dropped. And now, Mr. Holmes, you may well smile and accuse me of being silly, but I tell you I still cannot help but connect these deaths with the presence of this mysterious passenger."

"I presume you are referring to the deaths of a Dr. and Mrs. Ethan Russell?" said Holmes.

"Good God, you know all about the matter!" exclaimed Robinson, almost jumping out of his seat.

"Not *all* about the matter," replied Holmes. "I did know *something* about it, though less than I do now. I am a criminal investigator, Mr. Robinson, and it is my business to know about these things.

"I heard about this affair shortly after it happened," continued Holmes. "I formed my own theory at the time, and everything you have told me has only tended to confirm it. But there are still one or two points I should like to clear up. You wouldn't by any chance have any recollection or record of the chess position at the time when this passenger—let us call him Mr. M.—surprised you by telling his companion that there was a promoted piece on the board?"

"Why, yes, indeed," replied Robinson, reaching for his wallet. "My companion and I, after first commenting on the pair's incredible rudeness, were wondering how on earth this Mr. M. could know about the promotion. We spent close to an hour trying to figure out this mystery, but got nowhere. We then each made a diagram of the position obtaining at the time Mr. M. approached the game. I have kept this diagram with me ever since, and have several times studied it anew, but have not yet been able to solve it. Perhaps you can, Mr. Holmes?" he said, and he showed us the following diagram:

Black-14

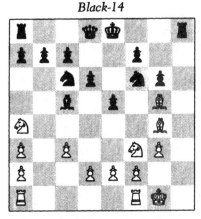

White-14

Holmes studied the position for some time, and shook his head. "I don't believe it *can* be deduced from this position that there is a promoted piece now on the board," said Holmes, "although it is trivial to see that a promotion must have taken place. But why does the promoted piece have to be on the board *now?*" puzzled Holmes. "I'm sure I can reconstruct the position in a manner in which this is not so."

Holmes studied the position a bit further. He then said, "Are you *absolutely* sure, Mr. Robinson, that Mr. M. did not see *any* move of this game?"

"Only the last one, which happened just after he came up," replied Robinson, "though I don't see how that could help."

"And what was the last move?" asked Holmes eagerly.

"I castled," replied Robinson. "I was playing White, you know."

"Oh!" said Holmes, "that puts an entirely different complexion on the whole situation! So now, given that White has just castled, I think I dimly see a possibility of proving that there must be a promoted piece on the board. Still, the problem is nontrivial, and I will need more time to check it. Would you have any objections to my making a copy of this position for my files?"

"Not at all," replied Robinson.

Holmes did this, and then said, "And now, Mr. Robinson, I have a question of particular importance: How long would you estimate that it took between the time you castled and the moment when Mr. M. told his companion about the promoted piece on the board?"

"Not long, Mr. Holmes; I would say about three minutes—certainly not more than four."

"Thank you, Mr. Robinson, this information may prove most helpful. And now if you gentlemen will excuse me, I should like to retire a bit and work on this problem."

Holmes walked off, and returned to us about a half-hour later. "Mr. Robinson," he said, "I hardly think you need worry about suffering from an overly active imagination. Your intuition was absolutely correct! The mysterious passenger of your singular experience happened to be one of the greatest criminal masterminds of the century. He also, incidentally, happened to be a professor."

"Holmes," I asked sometime later that afternoon when we were alone, "how were you so sure it was Moriarty? Of course, Robinson's description fitted him perfectly, but was there anything else?"

"Yes," replied Holmes. "I was particularly anxious to get the exact chess position for two reasons. In the first place, I wanted to know just how difficult this problem was. Well, Watson, I am certainly no novice at retrograde analysis, but it took me over twenty minutes to solve it. And here, this passenger solved it in three or four minutes! I can think of no one in the world who can think that fast other than either Moriarty or my brother, Mycroft. And, aside from the fact that Mycroft never takes trips, Robinson's description would hardly fit my brother," added Holmes with a laugh.

"Secondly, Watson, I was very curious to ascertain just why this passenger acted so peevishly when Robinson asked him whether he knew whether the promoted piece on the board was White or Black. Now, despite Moriarty's brilliant brain, he had the emotional maturity of a petulant child! He absolutely could not stand criticism, and would get into a cold fury whenever asked a question to which he did not know the answer. I'm sure had he known the answer, he would not have behaved so rudely, but would only too readily have flaunted it. Now, why did he *not* know whether the promoted piece was Black or White? Was it possible that the matter was indeterminate? Well, it turns out that this chess problem has the remarkable feature that although one can prove that there is a promoted piece on the board, it is impossible to deduce which color! This is the only problem of this type I have ever seen. So that's why Moriarty didn't know; he simply couldn't have!

"These facts, together with Robinson's rather remarkable description, establish beyond any reasonable doubt that it was Moriarty who was aboard that fateful voyage. The only thing about this entire affair about which I am still in the dark is the *motive* for this murder. What connection could there be between Moriarty and the Russells?

"And now," continued Holmes, "do let me show you the very beautiful solution to this singular chess problem."

SOME CHILLING
REMINISCENCES

MAY 18 (11:15 P.M.) We are now in the midst of a violent storm which came up quite suddenly about 7:30 P.M. Either this liner is remarkably stable, or there is surprisingly little turbulence, considering the violence of the rain and the magnitude of the thunder and lightning. Those of you who have ever experienced a storm on the high seas know what a dramatic and spectacular event it can be!

Most of the passengers have long since retired. Holmes and I turned in about 9:30, but neither of us could sleep. It was partly the storm, and perhaps even more the memory of this day's events, which hung heavily on both our minds. After a restless attempt at slumber, we both got up, lit our lamps, and continued discussing what had taken place. Somehow, the storm served as a most fitting background for our conversation.

"The worst of it, Watson, is that I had *some* evidence that Moriarty was about to take that passage. Had I been present, this tragedy might well have been averted. But Moriarty cleverly decoyed me by staging a glamorous and spectacular crime to cover his departure."

"Wasn't this about the time of the attempted Crown-jewel robbery?" I recalled, straining my memory.

"Exactly, Watson! I saved the Crown jewels, but at the cost of two human lives! I only wish it had been the other way round!"

We both sat silently and somewhat gloomily for some time, Holmes puffing away with more than usual energy at his pipe. "You know, Holmes," I said at last, "this is the second time that Moriarty has come up in connection with retrograde analysis. You recall how surprised I was the first time—the time you showed me that monochromatic problem which you associated with the story of the lion and the bear. I had no idea that retrograde analysis was amongst Moriarty's many talents."

"Oh, yes, indeed, Watson, Moriarty was one of the greatest retrograde analysts the world has known. Indeed, I regard his research on this subject as of even higher significance than his better-known work in mathematics. I have collected a dozen or so of his problems which I have in my files. When we get back home, Watson, do remind me to show them to you.

"In addition," continued Holmes, "on two separate occasions in my encounters with him, I received from him a threat by mail—each in the form of a chess problem."

"How singular!" I exclaimed.

"Yes, Watson, I think you might well publish these in your future chronicles.

"The first threat was in the form of the following diagram:

Black-11

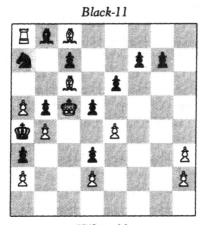

White-11

"Under it was the message:

HOLMES, YOU ARE PROVOKING ME BEYOND ENDUR-
ANCE! UNLESS YOU DESIST—AND I MEAN IMMEDI-
ATELY!—I WILL MATE YOU AS QUICKLY AS WHITE CAN
MATE BLACK. AND IT IS MY MOVE, HOLMES.
REMEMBER THAT!

"Well, Watson, the immediate question I had to solve
was this: In how many moves *can* White mate Black? The
solution nicely combines retrograde analysis with the more
usual forward play."

After Holmes showed me the solution of this problem, he
continued, "Actually, Watson, from a purely practical point
of view, this threat amounted to naught. Though a lovely
piece of chess bravado, I think he sent it more to unnerve
me than anything else. The second threat, however, was far
more serious, and, curiously, as it turned out, saved my life."

"Saved your life?" I asked in astonishment.

"Yes, Watson. This happened at a period when Moriarty
had *me* on the run! For two utterly exhausting weeks, I had
been dodging one attempt on my life after another. My
method was to disappear like an invisible phantom from one
spot of London to another. This took almost all my wits and
all my energy! Needless to say, I did not spend one night at
home during this harrowing period; I stayed each night at
the house of a different confidant. Well, one morning I
found at the establishment where I was staying a letter ad-
dressed to *me!* 'Good God,' I thought, 'if this is from
Moriarty, then he must have known I spent the night
here—so why has no attempt been made on my life?' With
feverish haste, I tore open the envelope, and sure enough,
the message *was* from Moriarty. There was the following
diagram:

Black-11

White-12
White king not shown

"And under it the message:

YOU CERTAINLY HAVE A REMARKABLE CAPACITY FOR
MAKING YOURSELF INVISIBLE, HOLMES.
NEVERTHELESS, I CAN MATE YOU IN ONE MOVE!

Holmes handed me the diagram, which I began to study. "Why, Holmes," I cried, "there is no White king shown on the board!"

"Of course not, Watson, that is the whole point! As Moriarty kindly informed me, the White king is *invisible*."

"The White king?" I asked, without yet comprehending the true nature of the situation.

"Oh, Watson," laughed Holmes, "why are you so literal-minded? Obviously in this position, Moriarty identified himself with the Black forces, and me with the White. So here I am the White king, who, though invisible, can be mated in one move."

"How did you deduce that?" I asked.

"Really now," laughed Holmes, more heartily than ever,

"you are sometimes too much! One need hardly *deduce* this, my boy; this part is intuitively as self-evident as can be.

"At any rate," he continued, "my immediate problem was to solve this mystery as quickly as possible. Could it be determined on which square the White king had to stand? Exhausted as I was, I strained every ounce of my being towards getting the solution. Yes, it turned out that by retrograde analysis, the king could be on only one square. And indeed there was then a mate in one!

"Then, Watson, out of the blue, the following chilling realization occurred to me—the very realization which, as it turned out, saved my life. If you could place a giant sixty-four-square grid over the whole of London, thus enclosing London in a huge chessboard, with the north side representing Black, then the house at which I planned to spend the coming night lies well within the very square on which must stand the White king! Whether this was sheer coincidence, or whether it was part of Moriarty's fiendish design, is something I am not sure of to this day. At any rate, I certainly took no chances, and abruptly changed my plans for the coming evening. Now, I have to tell you that the place I had planned to stay was blown up during the following night. Fortunately there were no occupants. But, I must confess, I found the experience unnerving."

"Indeed," I replied. "In many ways, Holmes, I find this one of the most curious adventures you have had. There is one thing, though, which dreadfully puzzles me. Why on earth did Moriarty ever send you that letter? It sounds to me more like a helpful warning than a threat!"

"Ah, Watson, that is one mystery I have never solved! I can think of only four possibilities. The first—which strikes me as the least likely—is that Moriarty simply underrated my ability to solve the problem. But I hardly think this does Moriarty justice. The second is that Moriarty, though confident that I could solve the chess problem simply as a chess problem, had no reason to believe that it would ever occur to me to think of the diagram as a crude map of London.

Indeed, as I told you, Watson, I have no idea how that thought *did* occur to me. The third possibility, of course, is that the correspondence of the crucial chess square to the region of London in which I had planned to stay was nothing more than a remarkable coincidence. However, none of these three possibilities can even remotely explain why I had not been done away with on the *previous* night! Now, the fourth possibility, though exceedingly bizarre, is the only one I can think of which would explain that as well, and hence is the only one I can take seriously.

"You know, Watson, that despite Moriarty's rare brilliance, I think that he was somewhat psychotic. With his deranged mind, is it so out of the question that he was having so much fun playing cat-and-mouse with me and seeing me so painfully on the run, that he merely wanted to prolong the agony? To add credence to this hypothesis, this whole matter occurred rather early in our encounters—a time not too long after our paths had first crossed. I had not yet obstructed him in any of his really *major* projects. Could it not be that he wanted to play the game with me a little longer—until such time as I was likely to become a really serious threat?"

A DISPLACED BISHOP

MAY 19. The storm abated at about 6:30 A.M. as suddenly as it appeared. Today is as beautiful a day as we have had.

We were on deck rather early this morning. We came across Robinson and his companion in the midst of another game of chess. He appeared in far better spirits than on the previous day.

Black-15

White-14

A White bishop stood between a3 and a4. As we stood watching, both sides successively castled. Holmes then correctly adjusted the bishop.

"I don't think anything you could do would surprise me

now," laughed Robinson. "Still, I would be curious to know how you knew. How did you?"

The solution is fairly easy.

A REMARKABLE
MONOCHROMATIC

MAY 22. Today we came across Lord Ashley and his wife at a game of chess. Lady Ashley is a great chess enthusiast, and her skill is at least equal to that of her husband.

"Ah, good morning, Holmes," said Lord Ashley at our approach. "We are in the midst of a most unusual game. No piece has moved from a white square to a black square nor from a black square to a white square."

"Ah, a monochromatic game," replied Holmes.

"Is that what you call it? Well, this is the only type of game Lady Ashley and I ever play. But this particular game, quite aside from being monochromatic, has been highly unusual."

Black-4

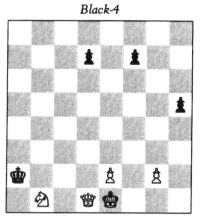

White-5

Holmes studied the position. "Whose move is it?" he inquired.

"It is mine," replied Lord Ashley, who was playing White.

"Then it is indeed a remarkable game," replied Holmes. "For one thing, I perceive a promotion has been made. For another, a pawn has been captured *en passant.*"

"Astounding!" said Lady Ashley. I too was astounded until Holmes explained the solution. And even after hearing the solution, I am still astounded.

LADY ASHLEY'S PROBLEM

MAY 22 (*some minutes later*) My wife, Ellen, has composed some charming monochromatic problems of her own," said Lord Ashley, with genuine pride. "Please let me show you one." He rearranged the board:

Black-4

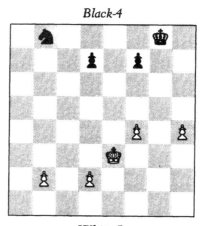

White-5

"In this game no piece has moved from a square of one color to a square of another. Also, the White king has moved but twice. The question is: Has the square h8 been reoccupied?"

A LITTLE MYSTIFICATION

MAY 23. Our last chess adventure aboard, though not dramatic, was a pleasant one.

Shortly before disembarking, Holmes and I were walking with Lord and Lady Ashley. We came across the following abandoned position:

Black-4

White-6

"Any clever things you can deduce about *this* game, Mr. Holmes?" asked Lady Ashley in her charmingly mischievous fashion.

"Only that you and Lord Ashley were not the players," Holmes cleverly replied.

How did Holmes know this? I can assure the reader that I have not withheld any clues.

ON MARSTON'S ISLAND

JUNE 15. Tomorrow we sail for home. This is our last day on Marston's Island, and quite a time it has been!

When we arrived, the sole inhabitants were Colonel Marston; his brother, Edward, who had come specially for the occasion; and the Colonel's recently hired native servant, Jal. We were all extremely eager to get on with the adventure and lost no time in settling ourselves upstairs in the old Captain's library. This library, numbering about eight thousand volumes, occupies the entire second storey. In one corner was a sofa, where the old fellow was reputed to have often slept after working late into the night. By the huge west window stood the Captain's desk—a four-by-ten-foot slab of oak resting on two old cabinets. The entire room represented the curious and intriguing combination of tastes of one who was both a scholar and a seafaring man. All the walls are lined with bookshelves, and the rest of the space is cluttered with maps, charts, personal manuscripts, brass telescopes, chronometers, and various nautical and surveying instruments.

Colonel Marston showed Holmes the map.*

"Can you make head or tail of it?" asked Marston.

"Not offhand," replied Holmes. "This, I'm afraid, will require luck as well as study.

"And now," he went on, "I would like to cloister myself in this library for a few days. I would even like to sleep on this sofa, if I may. I will need constant access to these books,

---

* The reader is referred to the next facing pages.—J.W.

and besides, I find this atmosphere most inspiring and conducive to thought."

We saw virtually nothing of Holmes for the next four days. He might well not have even eaten had not Marston had Jal bring him his meals. Then, on the morning of the fifth day, Holmes came down with a triumphant air as we were at breakfast.

"I believe I have solved the theoretical aspects of the matter, gentlemen. What is now called for is some highly accurate surveying."

"Perfect," cried Colonel Marston. "Surveying was my main occupation in the army, you know, and my great-grandfather's instruments, though quaintly old-fashioned, are as accurate as any I have seen."

We spent the next two days surveying the island. Towards nightfall, Holmes had mapped out a spot about forty feet square on the southwest corner of the island. "Tomorrow morning will tell," he said. "If my calculations are correct, the treasure should be found within this square."

Next morning at sunrise found the five of us with pick-axes, spades, and shovels. We started in the dead center of the forty-foot square. It did not take long before Jal's spade struck a chest. Ten minutes more and we had it unearthed. The box was about three feet long, two and a half feet broad, and two feet deep. It was bound by iron bands. I think it took us longer to break open the bands than to unearth the box—at least it seemed so in our state of excitement! But when we finally sprang the top open, lo and behold, the chest was empty except for one gold coin and a piece of parchment on which was written the message:

I KNEW YOU'D GET HERE SOONER OR LATER, HOLMES,
BUT AREN'T YOU A TRIFLE LATE??   M.

"Late by at least three years," said Holmes sadly.

I
B-11

W-12

A.K. *131—12, 3, 6;*
27—5, 14.
22—3, 16, 4, 18, 7, 14;
32—12, 2, 21?

II
B-13

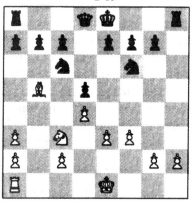

W-12

A. K. *63—3,4,5*
14—12, 18, 2, 21;
16—12, 4, 17, 5, 22?

III
B-8 or 9

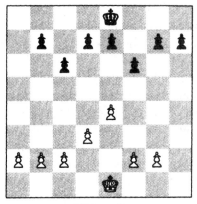

W-8 or 9

A.K. 71—3, 18, 9, 14, 22, 34, 60.
    28—2, 14, 12, 24, 16, 32, 27, 21.
    34—3, b6, 15, 22, *unknown.*
    42—3, 17, 9, 22, 12, 7?

Next morning, when Holmes came down to breakfast, he met three rather disconsolate associates. "Why so sad, gentlemen?" he asked in a voice which was surprisingly cheerful, considering the obvious disappointment of the episode.

"Oh," said Colonel Marston, "all our labor for naught! To think that all is lost!"

"Not necessarily *all,*" replied Holmes, more cheerfully than ever. "I'd say more likely *half* is lost."

"What are you saying?" cried Edward in astonishment.

"Well, gentlemen, I have just made a new discovery which leads me to believe that the Captain divided his treasure and buried it in *two* chests. If my calculations are correct, the other one should be somewhere on the beach on the north part of the east side of the island—not too far from this house, and surprisingly close to the water. Of course, we may again have been beaten to the punch, we can only try. So, if I may first get a bite of breakfast, some more surveying and excavating lies ahead."

Another day-and-a-half of surveying brought us to the spot. We had to dig more deeply than before, though the work went much faster, since we were digging in almost-pure sand. We unearthed a second chest, and when we finally got it open, we found to our relief that this time we had *not* been beaten to the punch. There before our eyes lay some hundred thousand pounds' worth of gold and jewels.

HOLMES
EXPLAINS IT

Three days later, after the great excitement had died down and after all the treasure had been transported to the house, sorted, and fairly well evaluated, the four of us were relaxing comfortably in the upstairs library.

"And now," said Colonel Marston, "tell us, how did you do it, Holmes? How did you decipher the message, and how did the chess diagrams locate the treasure?"

"Well," replied Holmes, "as to the cipher, it was, strictly speaking, a code rather than a cipher. With ingenuity, a cipher can be deciphered without any key; a code cannot. A code can be understood by the receiver only if he has the code book, or the equivalent. In this case, the numbers on the diagrams are clearly a reference to the words of some book. I realized at once that unless I could find the book, the task would be utterly hopeless—that is why I told you that this was going to be just as much a matter of luck as of skill. More specifically, if the book in question should have turned out not to be in this library, I should not have had the foggiest notion of how to find it."

"I still don't quite understand how the code works," said Colonel Marston.

"Well, look," said Holmes. "Look at the numbers: '131—12, 3, 6'—that means to turn to page 131 and then successively the twelfth word, third word, sixth word. Then '27—5, 14'—turn to page 27 and take the fifth and fourteenth words. And so on. . . .

"Actually," he continued, "it is more usual to take just one page for the whole message. This usually involves a

book with large, finely printed pages, like many editions of
the Bible, where there are an enormous number of words to
a page, and hence there will probably be all the words on
one page that one needs."

"So you tried over eight thousand volumes?" asked Edward incredulously.

"Hardly, my dear Mr. Marston," laughed Holmes, "although at first I was afraid that that grueling task might be
necessary. However, I figured that the book might well be
among the old Captain's four hundred volumes on chess,
since this subject was so close to his heart. And besides, I
hoped the opening letters, A.K., might be the initials of the
title. Fortunately this hope was realized. Here is the book."

The book was entitled *Arabian Knights,* by an unfamiliar
author, Nayllums Dnomyar. It was an old and beautifully
bound handwritten manuscript.

"This book is a veritable gold mine for any retrograde analyst," said Holmes. "It contains some of the finest retrograde problems I have ever seen, all clothed in delightfully
witty stories based on *Arabian Nights* characters. It is a
chess fantasy in which the pieces themselves are the characters—something like our own Lewis Carroll.

"Anyway, once I had the right book, the translation of
the message was immediate. As you can determine for yourselves, the messages read thus." Holmes laid a sheet of
paper before us.

I   NO WHITE PAWN HAS PROMOTED.
    ON WHAT SQUARE WAS THE OTHER WHITE BISHOP
    CAPTURED?

II   WHITE CAN CASTLE.
    ON WHAT SQUARE WAS THE WHITE KING'S ROOK
    CAPTURED?

III   WHITE GAVE BLACK ODDS OF BOTH KNIGHTS. NEI-
    THER KING HAS MOVED OR BEEN IN CHECK.
    ON H6 LIES AN UNKNOWN. WHERE WAS IT TWO
    MOVES AGO?

We all looked at the positions anew. "I still don't understand," said Marston, "how the solution of these problems locates the treasure."

"Well," replied Holmes, "each of the three problems involves locating a square. So we divide the island into sixty-four squares, as follows:

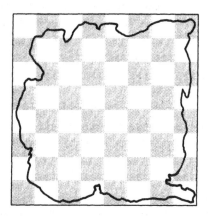

"The first problem, when solved, locates the square in which the treasure is buried. We then subdivide this square into sixty-four smaller squares, and locate one of them via the second problem. Repeating this process, using the third problem, we then obtain a square about forty feet on each side, since the island is approximately four miles square. All that remained was to solve the three chess problems."

"All that remained—a mere trifle for you, I suppose," said Edward humorously.

"Actually, they were not all that difficult," replied Holmes, "although the last one had me stumped for a while. Consider now the first problem. I started by observing that White is in check from the rook on a8. What was Black's last move?"

"Obviously with the king from a7 or a6," replied Colonel Marston.

"It couldn't have been from a7," said Edward Marston,

who also knew the rules of chess, "because the White rook on c7 would have had no way to check him."

"That's right," said his brother, "so it was from a6."

"Good," said Holmes. "This means that Black has just moved out of check from the bishop on f1. How did White administer this check?"

"It could have been by a White knight having moved from c4 to b6, where it was just captured by the Black king," I said, remembering a little exercise Holmes had shown me some months earlier.

"Or alternatively, a White rook from b5 to b6 could have given check," said Edward.

"Good tries, gentlemen," replied Holmes, "but they will not do. Both the missing White knight and White rook were captured earlier in the game by the pawn on h5 coming from f7. This must be so because the White pawn from c2 could not have captured enough pieces to get to the g-file, and we are given that it didn't promote."

"Why not enough pieces?" asked Colonel Marston. "The pawn from c2 had to make only four captures to get to g6, and Black is missing five pieces."

"You forget the pawn on g5," replied Holmes, "which must have captured two pieces to get from e2. This would involve just one capture too many."

"I see," replied the Colonel.

"So," continued Holmes, "the Black king did not capture a rook or a knight on his last move. Hence what was his last move, and what was White's move before that?"

"The position seems to me impossible!" I replied.

"Don't give up so easily," laughed Holmes. "You have forgotten, Watson, the old *en passant* trick! The Black king has just captured a White pawn on b6. White's move before that was with this pawn coming from c5 making a capture *en passant* of a Black pawn on b5—this gave check from the White bishop. Black's move before that was, of course, the pawn from b7 to b5, and White's move just before that was with the pawn from c4 to c5, putting Black in check from the bishop."

"Capital," said Edward.

"Now then," continued Holmes, "this means that two moves ago there was a Black pawn on b7. Hence the Black bishop on f7 has not had time to escape from c8. Hence this bishop is a promoted bishop."

"How unusual," cried Edward, who, though highly intelligent, had no previous experience with retrograde analysis.

"Not so unusual in retrograde analysis, as Watson can now well testify," laughed Holmes. "Anyway, the promoting Black pawn obviously came from a7. It made one capture on the b-file and promoted on b1. It must have made its capture on b2 behind the White pawn on b3."

"Just a moment," said Edward. "Why couldn't it be that the pawn on b3 really came from c2, in which case the Black pawn from a2 may have made its capture on some other square of the b-file?"

"Good question," replied Holmes. "However, if this was so, then the White pawn just captured by the Black king came originally from b2. It would have made one capture to get to the c-file and then made its capture *en passant.* Thus it would have made two captures. The pawn on b3 would have made one capture, and the pawn on g5 did make two. This totals five. But the Black bishop from c8 was captured on its own square without having moved. This makes one capture too many."

"How clever!" remarked Edward.

"So," continued Holmes, "this about solves the problem. The missing White bishop was not captured on g6 or h5, since these are white squares, and we know the Black king took a pawn on b6, hence it was the piece captured on b2. Thus b2 is the answer."

"What a delightful problem," said Colonel Marston, with evident family pride. "I wonder if the old Captain invented it himself?"

"I dare say he did," replied Holmes. "I hardly think he would risk using a problem which was known."

♟

"The second problem," continued Holmes, "is relatively easy. What piece was captured by the White pawn on a3? Not the bishop from f8, which never escaped from its home square; not the bishop from c8, which travels on white squares. The third missing Black piece is the pawn from h7. But this, of course, could not have gotten to a3. Hence the pawn from h7 has promoted. Now, this pawn crossed the second row (from White's side) neither on square f2 nor on d2 (since the White king never moved) but on e2. Thus the pawn made two captures to get to f3 (*after* the pawn from e2 moved to e3 and *before* the White pawn on f3 moved from f2), then one capture on e2, and then its fourth capture on either d1 or f1. Since the pawn on f3 did not move till *after* the Black pawn was on e2, the White king's rook could not have yet got out to be captured. Therefore the rook was the last White piece to fall, and hence was captured on d1 or f1. But as long as the Black pawn was on e2, this blocked the White rook from getting to d1. Hence the rook fell on f1."

♟

"The third problem," continued Holmes, "is an unusual one—actually two problems in one. First one has to find out what is the unknown; then one must determine where it was two moves ago."

"Just a minute," interrupted Edward. "Is it not possible that one might determine where the unknown was without knowing what it is?"

"Interesting question," replied Holmes. "I have not followed that line of investigation. All I know is that *I* had to find out first what the unknown is. And this is the problem which at first really had me stumped. There does not appear to be the slightest clue! Neither side is in check, and there is no immediate evidence of pawn captures. What clue is there? Where does one start?"

We all studied the position most carefully, but not a one of us could remotely fathom where to begin.

"I started on this problem late at night," said Holmes, "but got nowhere—I did not even know where to begin. Disappointed, I went to sleep. Then suddenly in the middle of the night, I jumped up in bed, and out of my sleep said, 'Of course, the Black king's bishop!' So the first clue lies with the Black king's bishop."

"The Black king's bishop is missing," I said. "It was captured on its home square without ever having moved."

"Of course," said Holmes. "The question, though, is what captured it?"

We then all saw the point, and the Colonel voiced it. "It couldn't have been a rook or the queen, since the Black king had never moved or been in check. Nor could it have been a bishop or pawn. Hence the bishop from f8 was captured by a White knight. But White gave Black odds of both knights. Hence the bishop was captured by a *promoted* White knight."

"Exactly," replied Holmes, "and the determination of the unknown then follows fairly easily. The promoting White pawn came from h2. It didn't promote via f7, since the Black king has never moved or been in check, hence it promoted via c7 after having made five captures. Also both Black bishops were captured on their home squares, so were not captured by this pawn. Also the Black king's rook was never captured by this pawn, since it was always confined to the squares f7, f8, g8, and h8. This then accounts for eight captures of Black pieces, so there can't be more than eight Black pieces now on the board. Hence the unknown on h6 must be White. The next question is *what* White piece."

"Maybe it is the promoted White knight," said Edward. "I think it would be rather droll if that were the case."

"It so happens it *is* the case," replied Holmes.

"How do you know?" I asked.

"Well, Watson, among the Black pieces captured by the White pawn from h2 was the Black pawn from a7, unless this pawn promoted. Well, it *must* have promoted, since it could never get on the diagonal from h2 to c7. Now, in the process of promoting, it never crossed the square d2, since the White king has never moved or been in check. Hence it first got to e2, having made four captures, and then made one more capture on d1 or f1. This totals five, and accounts for all missing White pieces, since there were only fourteen White pieces at the outset and nine White pieces are still on the board. Now, since the only Black pieces on the board are the king and pawns, the promoted Black piece—whatever it was—is no longer on the board. Therefore it was captured. Furthermore, it must have been one of the pieces captured by the White pawn from h2, otherwise there would be one too many captures of Black pieces. More specifically, the White pawn from h2 captured the Black queen, both knights, the queen's rook, and the promoted Black piece; the missing Black bishops and king's rook were never captured by the pawn."

"Why is it so significant that the promoted Black piece was one of the pieces captured by the White pawn?" inquired Colonel Marston.

"Because this means that the Black pawn promoted *before* the White pawn did. Hence the promoted White knight was *not* among the five White pieces captured by the Black pawn en route to promotion. And the promoted White knight was not captured by anything else, or there would be one too many White pieces now on the board. Therefore the White knight must be the unknown on h6."

"Very good," said Edward. "And now what about the second part of the problem? Where was the knight two moves ago?"

"Well," said Holmes in a delighted tone, "the answer shows that your great-grandfather must truly have had quite a sense of humor."

"He was certainly reputed to," replied Edward with a laugh.

"Well, the solution is this," said Holmes. "A Black piece must have been captured within the last two moves to avoid a Black retrograde stalemate."

"What is a retrograde stalemate?" asked Edward.

"A situation in which there is no possible last move," replied Holmes. "You see the whole point is that in this position, the Black pawn on c6 has been there for quite some time to allow the promoting pawn to cross c7, and therefore it has not moved for several moves—certainly more than two. So the last move that Black made must have been with the pawn on f6 coming from f7. And Black's move before that? It must have been with a piece no longer on the board. Thus Black's last move was with the pawn on f6, and right before that White must have captured a piece which was free to have moved previously. What piece could this be? Only the Black king's rook captured on f7 or g8 by the White knight. It couldn't have been f7 because the pawn was there. Hence it was g8. Thus before White's last move, the knight was on g8. What square did it come from to capture the rook on g8? Not f6, since it would be checking the king. Therefore it came from h6. In other words, put the Black pawn now on f6 back on f7 and put a Black rook on g8—this was the position two moves ago. The present position is then obtained by the following sequence:

1 N×R P–f6
2 N–h6

"Thus two moves ago, the knight was on h6—the same square it is on now!"

♟

We all agreed that the solution did indeed have a humorous twist—the idea that two moves ago the unknown was on the very same square.

"I see now," said Colonel Marston, "how the solution of these three problems located the treasure. But that was only the first chest. How did you find out about and locate the second chest?"

"By a mere fluke, Colonel Marston, a mere fluke. If it were not for extraordinarily good luck, the chest might have lain there for millions of years.

"It happened this way: When we came home Tuesday afternoon, after finding the first chest rifled, I went up to the library, wishing to forget the disappointing events of the day. I started browsing through the old Captain's chess books. By sheer luck I came across this message:

$$68—12,\ 2,\ 14,\ 22,$$
$$14—3,\ 15,\ 8,\ 12,\ 9.$$

If it had not been enclosed in a map of the island, I would probably have ignored it. Fortunately, the *Arabian Knights* was the code book for this message also. Decoded, the message reads:

FOR THE SECOND PORTION, USE THE PROBLEMS
IN REVERSE.

"That was surely explicit enough. We merely use the square of the *third* problem for the rough location—this brings us to the northeast portion of the island. Then we subdivide into sixty-four squares and use the second problem for the intermediate stage, and lastly use the first problem."

This about concludes our remarkable adventure on Marston's Island. We had made no prior agreement about our share of the treasure, if it were found. Both brothers were extremely generous, and pressed us to take far more than we were willing to accept. We finally compromised on a figure about midway between what they had in mind and what we thought was reasonable. Still, the Colonel was not satisfied.

"Don't you realize, Holmes, that if it were not for you, we would never have found the treasure at all?"

"I doubt that," replied Holmes with genuine modesty. "I really think any competent retrograde analyst could have done as well."

"Well, well," said Marston, "won't you at least accept some extra little souvenir of the occasion? Do any of these instruments or books strike your fancy?"

"Well," said Holmes a bit shyly, "one thing I would truly be delighted to have—the *Arabian Knights.*"

EPILOGUE

We have been home now for about seven months, and there are two aspects of the mystery which have been cleared up only this evening.

Colonel Marston is again on a visit to London, and has spent the evening with us in Baker Street reminiscing about the whole adventure.

"Who stole the first half of the treasure, I should like to know, and how did he ever locate it?" asked the Colonel.

"The identity of the thief is no mystery to me," replied Holmes, who then told him all about Moriarty. "And it is not at all out of the question that the old Captain made more than one copy of the map. Though how Moriarty ever came across another copy is as much a mystery to us as to you. I have never been able to trace any connection between Moriarty and your family."

We sat awhile in silent thought. Suddenly, out of the blue, Holmes asked, "You never by any chance knew the late Dr. Ethan Russell?"

"Why, of course!" cried the Colonel in utter astonishment. "His wife, Violet, was my first cousin—she was a Marston, and, in fact, at the time of her death, she, Edward, and I were the only surviving descendants of Captain Marston. She and her husband were en route to pay Edward a visit on the island; they had sent word to him earlier that they were coming, bringing a newly found document of great family importance. However, they both died aboard ship under quite mysterious circumstances."

"Well," said Holmes to me later in the evening, "this finally solves the mystery of the motive of the Russells' murder and the connection between Moriarty and the Marston family!"

One or two last things. The reader perhaps recalls that aboard ship, en route to Marston's Island, Holmes told me about some of Moriarty's retrograde problems which he had back home in his files. I append these problems and their solutions.

As to the *Arabian Knights* manuscript, much mystery surrounds its origin. The author's name appears to be a *nom de plume*, though we are not sure of this. The handwriting is in spots illegible, and we can only conjecture the meaning of some of the passages. Either the author never gave solutions to the problems, or the solutions were somehow separated from the rest of the manuscript and lost, so Holmes and I have to work out most of them. This and much other work remains to be done, but eventually we intend to publish this manuscript.

MORIARTY'S
PROBLEMS
and
SOLUTIONS

MORIARTY'S
PROBLEMS

M1*
Black 6

White 10

No captures have been
made in the last four
moves. It is White's
move. What was the
last move?

---

* The first four problems were all composed by Moriarty before the
age of nine—this one, which was Moriarty's first, was composed at the
age of seven. Simple as it is, it certainly shows remarkable precocity.

M2*
Black 2

White 3

Neither the White king nor queen has moved during the last five moves, nor has any piece been captured during that time. What was the last move?

M3
Black 9

White 3

No pawn has moved, nor has any piece been captured, in the last five moves. The Black king has been accidentally knocked off the board. On what square should he stand?

---

* For a child of eight, this is quite ingenious!

M4*
Black 11

White 14

The Black king has never moved nor been in check. White gave Black odds of a piece, not a pawn. What odds did White give?

M5†
Black 14

White 14

None of the royalty has yet moved.

A Prove that if there are no promoted White pieces on the board, then one of the four knights has moved.

B Prove also that if there are no promoted Black pieces on the board, then *two* of the knights have moved.

---

* This is the last problem Moriarty composed during his childhood, and is clearly the most mature.

† For some unknown reason, over twenty years elapsed between the composition of M4 and of M5; the latter was not composed till after Moriarty obtained his doctorate. This problem is a profound one, and it is interesting to note the professional manner in which the questions are phrased.

M6
Black 15

White 14

White can castle. Is the White queen on d2 original or promoted?

M7
Black 13

White 15

Both sides can castle. The unknown on c6 is obviously Black; we are given that it is not a rook.

A If it is a knight, then on what square was the missing Black queen captured?

B If it is the original queen, then on what square was the missing knight captured?

C If it is a promoted queen, then where was the knight captured?

M8
Black 14

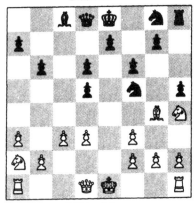

White 15

Given:

Black's first move was
  pawn to d5.
Knight on f5 has moved
  exactly three times.
Black queen, king, and
  king's rook have never
  moved at all.

Part I

A Prove that three of
    the pieces now on
    their home squares
    have moved.

B Has the pawn on h5
    moved once or twice?

C If the pawn h5 were
    on h6, would the posi-
    tion still be possible?

Part II—Suppose we remove the Black bishop from c8 and give
the following additional conditions:

  This bishop is somewhere on the board (and on a white
    square).

  It has never been on f7, nor has it crossed b7 or c6, and it did
    not move before the White bishop on g4 did. On what
    square is this bishop?

M9
Black 14 or 15

White 12 or 13

Neither king has moved.
On f2 stands a black or
white pawn. A White
knight stands on either
f3 or f4.

A What color is the
    pawn on f2?

B Does the White
    knight stand on f3 or
    f4?

149

M10*
Black 10

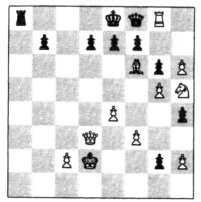

It is White's move. Can he mate in two?

White 11

---

* This is Moriarty's last problem—and one of his best. It was composed shortly before his death.

SOLUTIONS

THE MYSTERY OF THE INDIAN CHESS SET

Red is now in check, hence Green moved last. It remains to determine who moved first, which can be done by figuring whether an odd or even number of moves have been made.

The rook on b1 has made an odd number of moves; the other three rooks have each made an even number of moves (possibly zero). The Red knights have collectively made an odd number of moves, since they are on squares of the same color, and the Green knights have collectively made an even number of moves. The one king has made an even number of moves (possibly zero), and the other king an odd number. The bishops and pawns have never moved, and both queens were captured before they ever moved. So the grand totality is odd. Thus Green moved first. Hence Green is White and Red is Black.

ANOTHER QUESTION OF LOCATION

The White pawns have captured six pieces and the Black pawns eight. Among these eight must be a White pawn unless a White pawn promoted. Now, the White pawns on a4, b4, and c4 came from c2, d2, and e2, so if a White pawn was captured by a Black pawn, it must have come from f2, g2, or h2, but this is impossible; even from f2, it would have to make at least two captures to get to the d-file or beyond, which is one too many. Therefore, a White pawn did promote.

Now, the promoting White pawn could make at most one capture. Hence it came from h2 and promoted on g8 (since the Black king's rook has never moved), making its capture on g8 (from h7). Thus all seven missing Black pieces are accounted for. Also the Black pawn from h7 could not have been the piece captured by the promoting White pawn from h2, nor could it make enough captures to be captured by any of the pawns on a4, b4, and c4. Therefore this Black pawn must have promoted too. It could not have come straight down the h-file because of the promoting White pawn, which did not leave the h-file until after it was on the square h7. Therefore the Black pawn captured somewhere on the g-file (it couldn't have made more than one capture). So the ambiguously placed White pawn must be on f2, which allowed the Black pawn to cross to the square g2 (as well as allowing the White pawn on g2 to have been captured by the Black pawn).

HOLMES SETTLES A DISPUTE

The solution is indeed elementary. The point is that Black has not yet had time to move his king or either rook.

Before White's last move with the pawn from e2 to e4, he had nothing to move except the pawn from a2. Moving as slowly as possible, it must have made at least four moves to get to a6, then one move capturing the pawn on b7 before it was in turn captured by the bishop. So White made at most five moves with this pawn. Hence White has made a maximum of six moves altogether, and Black could not have made more than six moves. At the same time, Black has to have made at least six moves (at least one each with the queen, both bishops, both knights, and the pawn on e5). It follows that Black has made exactly six moves and has not yet moved his king or either rook; thus he is able to castle.

Since Black has made exactly six moves, and White has made no fewer moves than Black, then White has made at least six moves. But also White has made at most six moves. This means White also has made exactly six moves! Thus White and Black have each made exactly six moves, and it is now White's turn.

THE CASE OF THE DROPPED PAWN

Black is missing one rook, which was captured by the pawn on h3. Since the Black king has not yet moved, how did the rook get out to be captured? Only by the pawns on b6 and c6 cross-capturing to let it out—i.e., the pawn on b6 came from c7 and the pawn on c6 came from b7. Now, the Black rook could not get out until after the capture on c6, because as long as the pawn on c6 was on b7, the bishop on f5 was on c8 and confined the rook to the squares a8 and b8. Thus the capture on c6 was made first, then the bishop and rook both got out, and then came the capture on b6.

White is missing a bishop and a rook (the missing White pawn should really be on the board, although we don't yet know where). The missing White bishop traveled on black squares, hence was captured on b6. So the missing White rook was captured on c6. Thus the White rook was captured *before* the Black rook got out to be captured on h3, and therefore *before* the White pawn (coming from g2) made its capture on h3. But before the capture on h3, how could the White rook have gotten out since the White king has never moved? The only answer is that the White rook now on a1 is really the King's rook, and hence it was the White queen's rook which got captured on c6! The sequence was this: First the White queen's rook got captured on c6. Then the Black rook got out and got captured on h3. Then the rook from h1 got out and came around to a1.

We now see that it was the White queen's rook that was captured on c6, and we recall that the White king has never moved. Therefore the White pawn we are trying to locate cannot be on c2, since if it were, the White queen's rook could never have gotten out to be captured. Nor can it be on c3, since a cross-capture of this pawn and the one on b3 is not possible, because the only missing Black piece has been captured on h3. However, the White pawn we are trying to place must be on the c-file, because it was originally from c2, and had no Black pieces to capture to get onto a different file. Therefore it must be on either c4 or c7. It can't be on c7 because the Black pawn on c6 was there *before* the pawn on b6, hence there was always a Black pawn on c6 or c7, and the White pawn could never get around them. Thus the White pawn must stand on c4.

FROM WHERE?

The pawn on c3 could not have captured the Black queen's bishop (since this bishop traveled only on white squares), so either it captured the missing Black pawn or else this pawn promoted. If it did promote, then it was this promoted piece which was captured on c3, because this piece is not now on the board. Therefore the missing Black pawn was captured on c3 in either its original or its promoted form.

The missing Black pawn must be from d7, since if it were from e7 it would have had to make at least one capture to get either to c3 or to promote, and the pawn on e5 would have to have made another, which is impossible. So the missing pawn did come from d7. If it got captured on c3 without promoting, then it captured a White piece to get to the c-file. On the other hand, if it promoted, and then got captured on c3, it of course would have promoted *before* the capture on c3, hence the pawn on c3 was on d2 before the promotion, and the Black pawn must have promoted on e1 after capturing a White piece to get off the d-file. So in either case the Black pawn made a capture *before* the capture on c3. The capture was of the White bishop from f1 (since this is the only missing White piece). But before this, the pawn on e3 must have moved from e2 to let the bishop out.

The upshot of all this is that the pawn on e3 was on e3 *before* the pawn on c3 made its capture from d2. Therefore the White bishop on g5 got there from its home square c1 via the square f2, so White's last move with the pawn on f4 could not have been from f2. (Stated otherwise, if White's last move was the pawn on f4 from f2, then the bishop could never get from c1 to g5 because it would be hemmed in by the pawns on c3, e3, and f2.) Also, White did not just move the pawn on f4 from g3, as it would first have to have moved from f2, and this together with the pawn on c3 would have to have made three captures. Therefore White's last move was from f3.

DIFFICULT?

The missing Black bishop was clearly captured by a pawn on f3. It couldn't get out until the capture at c6 of one of the White rooks. Before the capture at f3, the White king's rook couldn't get out because the White king's bishop was in the way. So the White queen's rook was taken at c6. For it to get out, the pawns at b3 and c4 had to be in their present position (this is obvious for the pawn on c4, and as for the pawn on b3, the bishop on e5 couldn't have moved from c1 before the pawn moved from b2). So by the time the White king's bishop was freed by the capture on f3, the pawns at b3, c4, c6, and d7 were already there, so the White king's bishop couldn't get to where it is.

THOUGHTS OF A LOGICIAN

First, the solution of the problem of the knights and knaves. Since C contradicted B, then either B or C is a knight and the other a knave. If A is a knight, then there would in fact be two knights present, in which case A, being truthful, wouldn't say there was just one. On the other hand, if A is a knave, then there is in fact just one knight present, so A, being a knave, wouldn't state this true fact. Therefore A never said that there was one knight present. So B lied. Hence B is a knave and C is a knight.

Now for Fergusson's chess problem. If Black *cannot* castle, then White king to e6 does the trick, since whatever move Black then makes, pawn to g8 wins.

If Black *can* castle, then his last move was not with the king or rook, hence was with the pawn, moreover from e7 to e5. Then White can take him *en passant*. If Black then castles, to avoid mate by pawn to g8, White mates by pawn to b7.

Fergusson's whole point was that *either* Black can't castle *or* White can capture *en passant*, and there is no way of knowing which. In either case there is a mate in two, but a different one in each case. Thus, without knowledge of the past history of the game, there is no first move of White which one can point to and say, "This, and only this, move leads to a mate in two."

A QUESTION OF PROMOTION

The piece captured on c3 was not the Black queen's bishop, nor the pawn from h7, since it could not get to the c-file. Hence the pawn from h7 has promoted. It promoted on g1 after capturing exactly one piece—the White queen's bishop. The capture by the pawn on c3 had to occur before the bishop got out to be captured by the Black pawn, and so the capture on c3 occurred before the promotion. Thus the promoted Black piece is still on the board.

SHADES OF THE PAST

This is as complex a retrograde analysis as I have ever seen; in comparison, the problem discussed in "You Really Can't, You Know" is child's play!

Black is missing a bishop (from c8, which travels on white squares) and a pawn from g7 or h7. The Black piece captured by the White pawn on a3 was not the bishop, nor the pawn (which could never have gotten to the a-file), hence the missing Black pawn has promoted. If the Black piece captured on a3 was an *original* officer, then one of the Black officers now on the board must be promoted. Suppose, on the other hand, that it was the *promoted* Black piece which was captured on a3. This means that the promotion took place prior to capture and implies that there is a promoted *White* piece on the board, which must now be proved. Let's suppose the promoted Black piece was captured on a3. The first thing to observe is that the pawn on g3 came from g2, because if it came from h2 it captured on g3, which is a black square. But the pawn on a3 also captured on a black square. This is impossible, since one of the two missing Black pieces is the bishop traveling on white squares. So the pawn on g3 really did come from g2.

The next point to observe is that the promoting Black pawn did not promote on h1, since White has just castled. One consequence of this is that the Black pawn on g6 really came from g7, for if it came from h7, the promoting Black pawn came from g7 and would have had to make at least two captures to avoid promoting on h1. This totals three captures, which is one too many. Therefore the pawn on g6 really came from g7, and the promoting Black pawn came from h7, and must have made one capture on either g2 or g1. The crucial question now is, what White piece did this pawn capture? It could not have been a *promoted* piece, for suppose it was—then the White pawn from h2 promoted *before* the promoted pawn was captured by the Black pawn. How did it get past the Black pawn? The only possibility is that while the Black pawn was still on h7, the White pawn was on h6 and captured a piece on g7. But this is impossible, since a3 and g7 are both black squares, and one of the missing Black pieces is the bishop on white squares. Therefore the promoting Black pawn did not capture a promoted piece. And it did not capture the White pawn from h2, since it made its capture on g2 or g1. Therefore it captured an original White officer. If this officer was a rook, knight, or bishop, then one of the rooks, knights, or bishops now on the board must be promoted. But suppose the promoting Black pawn captured the queen. This capture occurred, remember, *before* the capture on a3 (because the Black pawn promoted *after* it made a capture, and the promoted Black piece then got captured on a3). So the White queen got out while the pawn on a3 was still on b2, and the pawn on c3 must have first moved to let the queen out (since White has just castled). Thus the pawn on c3 was where it is *before* the pawn on a3 came from b2. This means that the White bishop on g5 could never have gotten there from c1 (because it would be hemmed in by the pawns then on b2, c3, and d2), hence must be a promoted bishop.

To summarize, the Black pawn from h7 did promote. If this promoted piece is on the board, then there is a Black promoted piece on the board. If this promoted piece is not on the board, then it got captured on a3 after having captured an original White officer on g2 or g1. If this officer was not a queen, then one of the White officers on the board is obviously promoted. If this officer was a queen, then the bishop on g5 is promoted.

SOME CHILLING REMINISCENCES

The answer is that White can mate Black in only one move! The proof follows.

To begin with, the pawn on d3 came from h7 after having captured four of the five missing White pieces. Hence none of the other Black pawns have ever been off their present files, for if so much as one were, then at least two extra captures would have had to have been made. Thus the pawns on a3, b5, d5, and e6 have respectively come from a7, b7, d7, and e7, and have never made any captures.

Now, one of the White bishops on c6 and c8 is obviously promoted. The pawns on a5 and b4 have collectively come from b2 and c2 and have collectively made two captures (at least). The pawn on h3 came from g2 and has made one capture. This accounts for at least three captures. The promoting White pawn came from one of the squares f2 or e2, and the pawn on e4 came from the other. Suppose the promoting pawn came from e2. Then it must have made at least two captures to promote on a *white* square, and the pawn on e4 must have made another, which is one too many. Therefore the pawn on e4 came from e2 and the promoting pawn came from f2. This pawn went directly to f6, then made a capture on e7 and promoted on e8 (because otherwise it would have had to make three or more captures to promote on a white square).

Now we consider what must have been Black's last move. It was obviously not with the king, bishop, or knight, nor any of the pawns on a3, c7, f7, and g7. It was not with the pawn on d3, since had it just come from d4, it could never have come from h7. It could not be with the pawn on e6 from e7, or else the promoting White pawn never could have made its capture from f6 to e7. It could also not be with the pawn on d5 from d6, because it would have checked the White king, nor could it have just moved from d7, because the promoted White bishop could not then have left the square e8 on which it promoted! Therefore Black's last move was with the pawn on b5. It could not have just come from a6 (since it has never left its file), nor from b6, where it would have checked the King, hence it just came from b7. Therefore White can take him en *passant*, which mates Black.

Since it is Black's move, White is not now in check, hence the White king does not stand on any of the squares on the diagonal between the White bishop on h1 and the Black king. Therefore Black is in check from the White bishop. Obviously the check was discovered—i.e., some White piece moved from g2, f3, e4, or d5. It could not have been the pawn on h3 from g2, or else the White bishop could not have gotten to h1. Nor was it the pawn on g3. Nor could it have been any other piece except the pawn on d6 or the White king. Suppose it was the pawn on d6. Then it could not have come from d5, where it would have checked the king, but it could have come from e5 or c5 and captured a pawn on d5 *en passant*. However, it could not have come from e5, for it ultimately came from c3, hence would have made at least three captures, and these, together with the two captures made by the pawns on g3 and h3, would total five, which is one too many, since the bishop from f8 was captured on its own square. Thus if the pawn on d6 moved last, capturing a pawn *en passant*, it came from c5 and not e5. Before that, the captured Black pawn just came from d7, intercepting check from the bishop. How could White have given this first check? Only by the White king discovering it. (We might remark that the importance of showing that the White pawn came from c5 rather than e5 is this: If the pawn had come from e5, it could have discovered the first check by moving from e4. But on c5, this is not possible.)

We have thus shown that the White king either on the last move, or on the move immediately before that, has discovered the check to the Black king from the White bishop. From which of the four squares, d5, e4, f3, or g2, could the White king have come? Obviously not from d5. Also not from e4, for if it had, the bishop on d3 would have had to capture a piece in order to check the king, which together with the capture on a6 and g6 would total three captures on white squares, which is not possible since the White queen's bishop is among the three missing White pieces. Therefore the White king did not move from e4. It also did not move from f3, because it would have been in imaginary check from the queen and knight. Therefore it moved from g2. (It moved out of check from the queen and knight, but this check was not imaginary, since Black's last move could have been—indeed, must have been—a pawn from f2 capturing on e1 and promoting to a knight.) Thus the White king has moved from g2 and stands on a square which is not in check; this square must be g1. Thus the invisible White king stands on g1. Then knight to f3 mates White in one move.

A DISPLACED BISHOP

Clearly the missing Black pawn couldn't have been captured at b3, so it promoted. The promoting Black pawn never crossed the square d2, since the White king has not moved, hence the pawn must have first gotten to e2 and then made a capture on d1 or f1. This means that the promoting pawn came from e7, for if it came from d7, then it made two captures and the pawn on d6 came from d7 and made one capture. This totals three captures, but

White is missing only two pieces (a bishop and a pawn). Therefore the promoting Black pawn came from e7 and made a capture on d1 or f1. What did it capture? Clearly not a pawn. And the pawn from e2 couldn't have promoted, because it had no pieces to capture to get off its file and the Black king has never moved. Therefore the promoting Black pawn captured on d1 or f1 the White bishop traveling on white squares. Hence the ambiguously placed White bishop must travel on black squares. Hence it is on a3.

A REMARKABLE MONOCHROMATIC

The only way the Black king could have escaped from his home square e8 is by castling on the king's side and then coming out via h7. This means that Black's last move was not with the pawn at h5 from either h7 or g6, because it would then have hemmed in the Black king. Therefore Black's last move was with the king coming from b3. It just moved out of check from the White queen. How did the queen give this check? Not by having moved from c2, d3, or d5, where it would have checked the king. The only possibility is that the Black king has just captured a White rook on a2 which previously moved from c2, discovering check from the queen. So before the last move, there was a White rook on a2. Now, in a monochromatic game, an original White rook can never get to an even-numbered row, since it can move forward or backward only an even number of squares at a time. Therefore the White rook just captured on a2 was a promoted rook, which means a White pawn from a white square has promoted. For a pawn to reach the eighth square in a monochromatic game, it must make a minimum of four captures (assuming it moves two squares on its first move, rather than making a capture; otherwise it must make six captures). However, it could have captured only *three* pieces on white squares, namely the pawn from b7 (either in its original form, or the piece to which it may have promoted), the bishop from c8, and the rook from a8. All other missing Black pieces are from black squares (except, of course, the knight from g7, but this never moved). Therefore the promoting White pawn must have captured a piece on a black square! The only way this can happen is by making a capture *en passant*.

More specifically, what must have happened was this: The promoting White pawn came from a2 or c2, moved two squares on its first move, then captured a piece on b5 and then made a capture *en passant* on a6 or c6, then another capture on b7, and then its final capture on a8 or c8.

LADY ASHLEY'S PROBLEM

The only way the Black king could have escaped from e8 is by castling on the king's side. Now, unless the king subsequently moved to h7, the king's rook would have been confined to d8 and f8 and couldn't have gotten captured, for what could have captured it? Not the bishop from c1, not the king (which has moved only twice), not the rook from a1 (which can move only to an odd-numbered row), nor a pawn on black squares (since all four White pawns on black squares are still on the board), nor a promoted piece on black squares. Therefore after Black castled, the Black king moved to h7, the Black rook moved back to h8, the king then moved back to g8, and the rook was free to escape.

A LITTLE MYSTIFICATION

Lord Ashley said earlier that he and Lady Ashley played only monochromatic games, and Holmes could see from this position that it had not been a monochromatic game, and hence that it had not been played by the Ashleys. His proof is a clever one and nicely combines many features that we have experienced in the past monochromatics. The proof runs as follows.

Suppose the game was played monochromatically. Then we get the following contradiction. The Black queen, being on a white square, is obviously promoted. The promoting Black pawn came from d7 or h7 and made either six captures or four captures, depending whether on its first move it made a capture or moved two squares. Now, it couldn't have made six captures because White is missing only three pieces on white squares, and it couldn't have made more than one capture *en passant* of a pawn on a black square. So the promoted Black pawn has made four captures. It has captured the queen, king's bishop, a pawn from a black square *en passant*, and either the pawn from a2 or this pawn promoted. The latter would indeed be the case, for the following reasons.

The promoting Black pawn did not promote on b1 or d1, hence it promoted on f1 or h1 via g2. Now the White pawn from a2 could never have gotten in the path of the Black pawn from h7. And if it was the Black pawn from d7 which promoted, it first moved to d5, and thus made captures on e4, f3, and g2, so again the White pawn could not get in its path. Therefore the White pawn from a2 did promote. The only missing Black pieces on white squares eligible to be captured by the promoting White pawn from a2 are the queen's rook, queen's bishop, and the pawn from d7. This makes only three pieces. Therefore the White pawn made only four captures—one of them being a capture *en passant* of the pawn from c7. This also means, of course, that the White pawn moved two squares on its first move. And we might note incidentally that since the pawn from d7 was captured by the pawn from a2, it was the pawn from h7 that did the promoting.

Now, the pawn from d7 got captured on the diagonal from a4 to d7. It could not have made any capture (since all missing White pieces on white squares have been captured by the pawn from h7), hence it got captured on its own square d7. Hence the bishop from c8 also got captured on its own square. This raises the problem of where the Black rook from a8 got captured by the pawn from a2. Not on c8 or d7 (as we have just seen). Not on c6 (for the White pawn made a capture *en passant* on that square). Not a4 (for the pawn moved from a2 to a4). This leaves only the square b5. But the rook from a8 could never get to b5 in monochromatic chess. Therefore this position, in a monochromatic game, is impossible.

SOLUTIONS TO MORIARTY'S PROBLEMS

M1

To see what the position was four moves ago, move the Black queen to e4, the knight on e1 to f3, the Black bishop to e1, and the White bishop on c8 to h3. The following sequence of moves brought the game to the present position: (1) bishop to c8, check; (2) bishop to h4, check; (3) knight to e1, check; (4) queen to g4.

This is the only way the present position could have arisen, so Black's last move was with the queen from e4 to g4. (Try any other last move, and you will find it impossible to play back three more moves.)

M2

Put the Black pawn on a7, the Black king on g8, remove the White bishop, and put a White pawn on d5; this was the position five moves ago. The following sequence of moves brought the game to its present position:

	White	Black
1	P–d6	K–h8
2	P–d7	P–a6
3	P–d8 = B	P–a5
4	B–g5	P–a4
5	B–h6	

M3

The only way to avoid a retrograde stalemate for White is by placing the Black king on c8. Black's last move was with the rook from d8, White's move before that was with his king from g8, and Black's move before that was to castle.

M4

White had originally only fifteen pieces including the pawn from e2 (which was not given as odds). Either it was the piece captured on f6 or else it promoted. But the pawn couldn't have been captured on f6, because it would have to have made a capture to get to the f-file, and before the Black pawn made its capture, no missing Black piece could have gotten out to be captured. Therefore the White pawn promoted. It couldn't have promoted until *after* the Black pawn made its capture, hence an original White officer was captured on f6 and the promoted White piece is now on the board.

The pawn promoted on either d8 or f8 and hence it did not promote to a rook because it would have checked the Black king. It did not promote to a bishop, since the bishop on c1 is obviously original and the bishop on g2 is on a white square. Hence it promoted to a knight, and the two missing original White pieces are a knight and a queen, one of which was given as odds and the other captured on f6. Since the Black king has never been in check, it must have been the queen which was captured on f6. Therefore White gave Black odds of a knight.

M5

What was captured by the pawn on a6? Not the White queen's bishop (which travels on black squares), nor the missing White pawn, since it could not have made enough captures to get there. Hence the White pawn has promoted. Now, we are given that all the White officers on the board are original. So both original rooks, both original knights, and the original queen are present, hence were not captured on a6. And, of course, the White queen's bishop was not captured on a6. Therefore it was the *promoted* White piece which was captured on a6. Hence the promotion of the White pawn took place *before* the capture on a6. This means that prior to the promotion, the pawn on a6 was still on b7.

The promoting White pawn obviously came from d2 (because had it come from e2, it would have to have made at least two captures to get to the c-file or g-file, and the pawn on e4 would have to have made one, and only two Black pieces are missing). The pawn from d2 made one capture to get to the c-file and then either promoted on c8 or else made an additional capture on b8 (it couldn't have captured the Black pawn then on b7). Now, if it promoted on c8, then (since the pawn on a6 was at the time on b7) the original bishop from c8 could not yet have gotten out, hence was captured on its own square, which means that the bishop on a4 must be promoted. On the other hand, suppose the White pawn promoted on b8. Then it must have made two captures. Now the pawn from g7 could never have made enough captures to get even to the c-file, hence it must have promoted. Thus regardless of whether the White pawn promoted on c8 or b8, the Black pawn from g7 has promoted. It must have captured the White queen's bishop on h2 while the pawn on g3 was still on g2, then it promoted on h1. This means that the rook on h1 must have been somewhere else while the pawn on g3 was on g2 and the capturing Black pawn was on h2. It follows that the rook on h1 must have moved via g1, hence the knight on g1 must have moved. Thus the knight on g1 has moved, which proves the first part of the problem.

As for the second part, suppose that it is also the case that there are no promoted Black pieces on the board. Then in particular, the bishop on a4 is original. This implies that the White pawn from d2 could not have promoted on c8, hence promoted on b8, so the knight on b8 has also moved.

M6

The piece captured on b3 by the pawn was a Black rook. For it to get out, either the pair of pawns on a6 and b6 or the pair on d6 and e6 has cross-captured. Whichever pair did the cross-capturing, the following must hold: On the white square (a6 or e6) was captured the White king's bishop and on the black square (b6 or d6) was captured the missing White pawn, or else this pawn promoted. Now the missing White pawn is from a2 or c2. Also it has never captured any piece to get off its file, since the only missing Black piece has been captured by the pawn on b3. This implies that the missing White pawn is not from c2, since the pawn from c2 could neither get to b6 or d6 nor promote on the c-file because of the pawn on c7. Therefore the missing White pawn is from a2. Again it could not have gotten to b6 (nor, of course, to d6), hence it promoted. And, of course, it promoted on a8 without even having left the a-file. This means that the pair of Black pawns on a6 and b6 is the one which did the cross-capturing. Furthermore, the pawn on b6 made its capture while the pawn on a6 was still on b7, then the White pawn got to a7 or a8 and then the Black pawn from b7 made its capture of the White king's bishop on a6. The important thing now is that the capture on b6 occurred *before* the promotion, hence was of an original piece, so the promoted White piece is now on the board. What piece could it be? Not a knight, because a promoted knight could never have escaped from a8, since the pawn on b6 was there *before* the promotion. Not a bishop, because the White bishop now on the board is on a black square. Not a rook, because if the rook on a1 is promoted then White can't castle, and if the rook on g1 is promoted then the White king must have moved to let it in, and again White can't castle. Therefore the promoted piece must be the White queen.

M7

This complex solution may be facilitated by dividing it into steps:

1 Aside from the Black queen's rook, which got captured on a7, a8, or b8, Black is missing two pieces eligible to be captured on c3 and h3. One of them is the queen or a knight and the other is the pawn from g7. Now this pawn had no pieces to capture to get captured at c3 or h3 (since the only missing White piece— the queen—was captured at b6), hence this pawn has promoted. Also it promoted *after* the capture on h3 (since it couldn't get off its file).

2 The promoted pawn is either now on the board or it got captured at c3 or h3. It couldn't have been captured at h3 (because this capture occurred prior to the promotion). Hence it is either on the board or was captured at c3.

3 In either case, the promoted pawn did leave the square g1 (on which it promoted). Hence it could not have promoted to a knight, because the pawn on h3 was there prior to the promotion and the knight couldn't have escaped without checking the White king and making it move. Thus Black did not promote to a knight.

4 Since the White queen was captured at b6, the White pawn on c3 made its capture first to let the White queen out. Therefore the capture on c3 occurred prior to the capture at b6. This means that the original Black queen could *not* have been captured at c3, since it couldn't have gotten out till after the capture at b6, which in turn occurred after the capture at c3. So either the original queen stands on c6 or it got captured at h3.

Using these facts, we can go on as follows:

A Suppose the unknown on c6 is a knight. Then the original Black queen got captured on h3.

B Suppose the unknown is a promoted queen. Then the original queen was captured on h3, so the missing knight was captured on c3.

C Suppose the unknown is the original queen. There are now two possibilities: The promoted Black piece is either still on the board, or else it was captured on c3 (see Step 2). If the latter, then the missing knight was captured on h3. Suppose the former. Then the promoted piece is not a knight (see Step 3). It is also not a rook, since Black can castle. Hence it must be the bishop on g7. So the original Black king's bishop must have been captured on c3 or h3, and since h3 is not possible, it must have been on c3. Hence again it was the knight that was captured on h3.

To summarize:

A If c6 is a knight, then the missing Black queen was captured on h3.

B If c6 is an original queen, the knight was captured on h3.

C If c6 is a promoted queen, the knight was captured on c3.

M8

The Black queen's knight moved b8-c6-d4-f5. The pawn on d6 came from c7, capturing the White queen's bishop. The Black king's bishop must have been captured on its own square, and the Black queen's rook was captured at f3. For the Black queen's rook to get out, the capture of the White queen's bishop at d6 had to have happened. And before this, the pawn at d3 had to have moved there to let out the White queen's bishop. Therefore the pawns at d3 and, of course, f3 were there before the White king's bishop left its home square f1. Also the pawn on d5 was there before the White king's bishop moved (since that was Black's first move). Lastly, the knight on f5 was there before the White king's bishop moved, for the following reason: For the rook from a8 to get to f3 to be captured, it had to traverse the squares b8, c6, and d4—the very squares previously occupied by the knight on f5. The knight had to move ahead of the rook, so to speak, hence reached f5 before the rook reached f3, which in turn was before the White king's bishop was released. Therefore, we have proved that the pieces on d3, d6, f6, and f3 were all there before the White king's bishop moved, and have been there ever since. How then did the White king's bishop get from f1 to g4? Only by the following circuitous route: First it went to e2, then d1, then on or via b3, then on or via b5 (via a4 or c4), then on or via b7 (via a6 or c6), then on c8, e6, g8, h7, g6, h5, and g4. Thus the squares d1, c8, and g8 have all been traversed by the White king's bishop, hence the White queen, Black queen's bishop, and Black king's knight have all moved. Thus three pieces now on their home squares have moved. This answers the first question.

The pawn on h5 must have been on h6 when the White king's bishop moved from g6 to h5. Therefore this pawn has moved twice.

If the pawn on h5 were placed back at h6, the position would be impossible, because there would be no way for the Black king's knight to get back to g8 after the White king's bishop passed through, since the pawn at f6 and, of course, the pawn at e7 were already there.

In the last part of the problem, the White king's bishop had to follow the same path as before, and since the Black queen's bishop didn't move until the White king's bishop did, the additional constraints require that the Black queen's bishop precede the White king's bishop on its tour, so it must now be on h3.

M9

To begin with, the bishop on a5 must be promoted, since the Black king's bishop was captured on its home square f8 (the pawns on e7 and g7 have never moved). Hence a Black pawn has promoted. It must have come from f7 or h7 and promoted on the black square g1.

A Suppose the pawn on f2 is White. Then White is missing three pieces and Black is missing two. Now, the pawn on g3 could not have come from h2 or the bishop on a5 never could have escaped from g1. So the pawn on g3 is from g2, and it moved from g2 only *after* the Black bishop escaped from g1. Therefore the White pawn was always on g2 before the Black pawn promoted. Therefore the promoting Black pawn could not have come from f7, so it came from h7 and made its capture on the very square g1. Thus there have been captures on g1 and b6, both black squares. The White king's bishop was not captured on either of these squares, hence the White queen's bishop was captured on one of them and the White pawn from h2 was either captured on the other or else it promoted. It couldn't have been captured on b6 or g1, hence it promoted. Therefore both the Black pawn from h7 and the White pawn from h2 have promoted. Now, either the White pawn made two captures to get around the Black pawn or the Black pawn made two captures to get around the White pawn and in addition made a third capture on g1. The latter is impossible, since these three captures together with the capture on b6 are one too many, and the former is also impossible, since the two captures together with the capture of the Black king's bishop on f8 are one too many. Therefore the assumption that the pawn on f2 is White leads to a contradiction. Therefore the pawn on f2 is Black.

B Since the pawn on f2 is Black, then White is in check, hence this Black pawn has just moved. Now suppose the White knight is on f3 rather than f4. Then the Black pawn on f2 just came from e3, capturing a piece on f2, and it must then have previously captured another piece to get from f7 to the e-file. Hence the Black pawn on f2 has made two captures, the pawn on b6 one, and the promoting pawn from h2 one (though possibly on g2 rather than g1, since now the promoted Black bishop could have left g1 via f2). At any rate, this accounts for the captures of all four missing White pieces, and the pawn from h2 could not have been captured at g1 or g2, nor by the pawn from f7, hence must have promoted. Again we have the problem of how either of the promoting pawns from h2 and h7 got past the other. The White pawn from h2 has not made any captures at all (since Black now has fifteen pieces on the board, and the Black king's bishop was captured on its own square), hence it could not get past the Black pawn. So the Black pawn got past the White pawn. Either it made one capture on g2, while the White pawn was still at h2, or it made two captures to get around the White pawn somewhere on the h-file, and then made another one on g2 or g1. These three, together with the capture on b6, and the two captures by the pawn on f2, are two too many. Hence the former holds—i.e., the Black pawn captured on g2 while the White pawn was still on h2. Hence the White pawn on g3 was there *before* the Black pawn promoted to a bishop on g1. Hence the promoted bishop on g1 must have escaped to a5 via the square f2. This means that the White piece just captured by the Black pawn on f2 was *not* a White pawn, for if it were, the Black bishop from g1 could never have escaped! This now raises the following problem with the White pawn originally from f2: This pawn has made no captures (since the only capture of a Black piece occurred on f8), hence it was not captured by the pawn from f7 on the e-file, nor on f2 (as we have just seen), nor was it captured on b6 nor on g2 or g1, nor did it promote (since it could not have passed f7 without making the Black king move). So the White pawn from f2 neither promoted nor was among the four missing White pieces accounted for by the four captures made by the pawns from a7, f7, and h7. This is impossible. Hence the assumption that the White knight stands on f3 leads to a contradiction. Therefore it stands on f4 (and now the whole difficulty is avoided by the fact that the pawn on f2 could have come from f7 without making any captures at all).

M10

Yes, White can mate in two moves. First, we must prove that Black cannot castle.

Observe that the pawn on g2 came from c7 making (at least) four captures. Therefore the pawns on g6 and h4 have respectively come from g7 and h7, since if it had been the other way around, two more captures would have had to be made, which are one too many, since White is missing only five pieces. Therefore Black's last move was *not* with the pawn on g6 coming from h7. So what was Black's last move? Assuming it was not with the king or rook (in which case Black obviously can't castle), it must have been one of the following four:

A   The pawn on g6 from g7.

B   The pawn on g2 from g3.

C   The queen on f8 from g7 *without* making a capture on f8 (which implies that the White rook has previously captured a piece on g8 in order to check Black).

D   The queen on f8 from g7, but making a capture on f8.

Before treating these cases separately, some general observations are in order.

One of the four White pawns on e4, f3, g5, and h6 has come from d2, and the four pawns have collectively captured four Black pieces. The fifth of the six missing Black pieces is the bishop from f8, which got captured on its own square (because of the pawns on b7 and d7). Hence neither of the two missing White pawns from a2 and b2 has made more than one capture. Now, White is missing these two pawns and three officers. Four White pieces were captured by the pawn on g2 coming from c6. Neither of the two White pawns from a2 and b2 could have been captured by this Black pawn, because even the one from b2 would have to make at least two captures to get even as far right as the d-file. Therefore at least one of these two White pawns has promoted. So, the two things to remember are that at least one of the pawns from a2 and b2 has promoted; and that the pawns from a2 and b2 have not made more than one capture between them.

Now we return to the four cases (A, B, C, D) of Black's last possible move. As will be seen, cases A and C sort of belong together—as do cases B and D.

Let us first consider perhaps the simplest case—case C. In this case, White has on his last move captured a Black piece on g8.

This means another capture of a Black piece, hence neither pawn from a2 and b2 has made any capture at all. Likewise, if Black's last move was with the pawn on g6 from g7 (case A), then the Black king's bishop was captured on its own square, which again means another Black piece captured not by one of the four White pawns on e4, f3, g5, and h6, so again neither of the White pawns from a2 and b2 had any piece to capture. Thus, in either case A or case C, neither pawn from a2 and b2 has made any capture. Yet one of them promoted. It couldn't have been the one from b2 (because of the Black pawn at b7), hence it was the one from a2. So the pawn from a2 promoted without making any captures, hence must have promoted on the Black queen's rook's square a8, hence the rook has moved, so Black can't castle.

Now, cases B and D. Let us first consider D. The queen has just captured a *White* piece on f8. This, together with the four White pieces captured by the pawn on g2, accounts for *all* five missing White pieces, hence *both* pawns from a2 and b2 must have promoted (to provide enough White pieces to be captured by the Black pawn on g2). Likewise, if Black's last move was with the pawn at g2 from g3 (case B), then this pawn made all of its four captures on the *black* squares on the diagonal from c6 to g3; hence the White king's bishop got captured separately, and so both pawns from a2 and b2 had to promote (to provide officers to be captured by the Black pawn). Therefore, in either case B or case D, *both* pawns from a2 and b2 have promoted. Yet the two together have made no more than one capture. Now the pawn from b2 had to make a capture to promote, hence the one from a2 made no capture, hence promoted on a1, so again the Black queen's rook has moved.

This proves that Black cannot castle. White now mates in two by playing queen to d6. If the pawn takes the queen, then White mates by knight to g7. If Black moves the king to d8, White mates by the rook taking the queen on f8. If the pawn on e7 moves to e6, queen to e7 mates. If Black makes any other move, White mates by the queen taking the pawn on e7.

Other books by Raymond M. Smullyan available from Dover

Alice in Puzzle-Land
A Carrollian Tale for Children Under Eighty

First-Order Logic

King Arthur in Search of His Dog and Other Curious Puzzles

Satan, Cantor and Infinity
Mind-Boggling Puzzles

Set Theory and the Continuum Problem

The Lady or the Tiger?
and Other Logic Puzzles

What Is the Name of This Book?
The Riddle of Dracula and Other Logical Puzzles